Introduction to the Theory and Practice of Sampling

Introduction to the Theory and Practice of Sampling

Kim H. Esbensen

with contributions from Claas Wagner, Pentti Minkkinen, Claudia Paoletti,
Karin Engström, Martin Lischka and Jørgen Riis Pedersen

IMP**Open**

Published by:
IM Publications Open LLP, 6 Charlton Mill, Charlton, Chichester,
West Sussex PO18 0HY, UK
Tel: +44-1243-811334, Fax: +44-1243-811711,
E-mail: info@impopen.com, Web: impopen.com

ISBN: 978-1-906715-29-8

British Library Cataloguing-in-Publication Data
A catalogue record for this book is available from the British Library

© IM Publications Open LLP 2020

All rights reserved. No part of this publication may be reproduced, stored or transmitted, in any form or by any means, except with the prior permission in writing of the publishers.

Printed and bound in the UK by Hobbs the Printers Ltd

Contents

Foreword xi

1 **Theory of Sampling (TOS)—the missing link before analysis** 1
    1.1 A framework for representative sampling 5
    1.2 What comes before analysis—the TOS! 7
    1.3 What this book promises... 9
    1.4 References 10

2 **Theory of Sampling (TOS)—fundamental definitions and concepts** 11
    2.1 Lot dimensionality 12
    2.2 Sampling terminology—the tower of Babel 15
    2.3 References 19

3 **Heterogeneity—the root of all evil (part 1)** 21
    3.1 Introduction to the concept of sampling errors (excerpt from DS3077) 22
    3.2 Heterogeneity—the basics 25
    3.3 Materials, sampling targets and lots 25
    3.4 Homogeneity–heterogeneity 28
    3.5 Scale 30
    3.6 Heterogeneity vs sampling 31
    3.7 References 32

4 **Heterogeneity—the root of all evil (part 2)** 33
    4.1 Introduction 33
    4.2 Constitutional Heterogeneity (CH) 36
    4.3 Distributional Heterogeneity (DH) 37
    4.4 Heterogeneity vs practical sampling 41
    4.5 "Structured heterogeneity" 46
    4.6 The fundamental insight on how to counteract heterogeneity 48
    4.7 References 49

5 **"Sampling—is not gambling"** 51
    5.1 Introduction 51
    5.2 Enough analogy 52

6 **Pierre Gy's key concept of sampling errors** 57
    6.1 Rational understanding of heterogeneity and appropriate sampling 57
    6.2 Although complex, there is hope 62
    6.3 How to sample representatively: the TOS 64
    6.4 References 66

7 Composite sampling I: the Fundamental Sampling Principle ... 71
7.1 WHAT TO DO with all this heterogeneity? ... 71

8 Composite sampling II: lot dimensionality transformation ... 83
8.1 1-D lots: *conveniently* elongated lots ... 83
8.2 Process sampling ... 85
8.3 Process sampling generalised ... 86
8.4 Q ... 87
8.5 Lot dimensionality transformation (LDT) ... 91
8.6 References ... 91

9 Sampling quality assessment: the replication experiment ... 93
9.1 Background ... 93
9.2 Clarification ... 95
9.3 Quantifying total empirical variability—the replication experiment ... 100
9.4 Relative sampling variability ... 101
9.5 Notes and references ... 109

10 Sampling quality criteria ... 111
10.1 Sampling quality criteria ... 111
10.2 First SQC component—definition of analyte(s) ... 112
10.3 Second SQC component—delineating the decision unit (DU) ... 112
10.4 Third SQC component—inference and confidence ... 113
10.5 Perspectives ... 115
10.6 Summary ... 117
10.7 References ... 117

11 There are standards—and there is *the* standard ... 119
11.1 First light ... 119
11.2 Analysis of sampling standards for solid biofuels ... 121
11.3 Analysis of grain sampling guide ... 123
11.4 Sampling for GMO risk assessment ... 125
11.5 Examples of too glib recommendations ... 125
11.6 TOS competence is crucial ... 127
11.7 Que faire? ... 129
11.8 DS 3077 Horizontal—a new standard for representative sampling. Design, history and acknowledgements ... 129
11.9 Chapter references ... 138

12 Spear sampling: a bane at all scales ... 141
12.1 Introduction ... 141
12.2 Spear sampling—at all scales ... 143
12.3 Not always bad—there is hope ... 147
12.4 Conclusions ... 148
12.5 References ... 150

13 Into the laboratory... the TOS still reigns supreme 151
13.1 Representative sampling—a scale invariant endeavour 151
13.2 Size does not matter—only heterogeneity, and how to counteract it 155
13.3 And there is more to be done in the lab 156
13.4 Further reading 159

14 Representative mass reduction in the laboratory: riffle splitting galore 161
14.1 Introduction... 161
14.2 Riffle splitting 162
14.3 Automation—enter the rotary divider 169
14.4 Benchmark study.. 171
14.5 The ultimate method/equipment ranking for the laboratory 173
14.6 Conclusions 175
14.7 References 176

15 Introduction to process sampling 179
15.1 Lot dimensionality: ease of practical sampling.......................... 179
15.2 Lot dimensionality transformation 183
15.3 Process sampling.. 184
15.4 1-D lot heterogeneity 187
15.5 Variographic analysis: a first brief 188
15.6 Interpretation of variograms....................................... 189
15.7 References 194

16 Process sampling: the importance of correct increment extraction............. 195
16.1 Moving, or static, 1-D lots: increment cutting must be TOS-correct 195
16.2 "Sooner or later".. 200
16.3 References 201

17 The variographic experiment 203
17.1 The variogram 204
17.2 References 211

18 Experimental validation of a primary sampling system for iron ore pellets 215
18.1 Introduction: status of current ISO standards 215
18.2 Fundamental Sampling Principle and basic requirements for iron ore sampling systems . 216
18.3 Principles and general requirements for checking sampling bias 218
18.4 Validation experiment 220
18.5 Experimental results... 220
18.6 Discussion 222
18.7 References 223

19 Industrial variographic analysis for continuous sampling system validation ... 225
19.1 Variographic analysis 225
19.2 Continuous control of sampling systems 225

19.3	9–12.5 mm size fraction of iron ore pellets	226
19.4	Specific surface area of magnetite slurry	228
19.5	Iron grade in magnetite slurry	231
19.6	Conclusions	232
19.7	Acknowledgements	233
19.8	References	233

20 Theory of Sampling (TOS): *pro et contra* 235
20.1	A powerful case for the TOS in trade and commerce	236
20.2	Cases against the TOS (science, technology, commerce, trade)	240
20.3	Important reading with which to catch the attention of newcomers to the TOS	245

21 Following the TOS will save you *a lot* of money (pun intended) 247
21.1	Case 1: Always mind analysis	248
21.2	Case 2: Saving a client from a wrong, expensive investment	250
21.3	Case 3: The hidden costs—profit gained by using the TOS	254
21.4	Case 4: The cost of assuming standard normality for serial data	255
21.5	Lessons learned	259
21.6	References	260

22 A tale of two laboratories I: the challenge 263
22.1	Introduction (scientific, technological)	264
22.2	There is analysis… and there is analysis+	265
22.3	The core issue	267
22.4	The crux of the matter	268
22.5	The complete argument	271
22.6	The meaning of it all	272
22.7	Inside and outside the complacent four walls of the analytical laboratory	275
22.8	"One fine day"…	276
22.9	The really important aspect: costs or gains	277
22.10	What in the world?	278
22.11	References	279

23 A tale of two laboratories II: resolution 281
23.1	Epiphany interpretation I: knowingly closing one's eyes or not?	281
23.2	Epiphany interpretation II: the economic dilemma	283
23.3	Epiphany interpretation III: the moral resolution	285
23.4	Laboratory B's new vision and mission	287
23.5	Can this really lead to increased commercial success?	288
23.6	Acknowledgements	289
23.7	References	289

24 Sampling commitment—and what it takes 291
24.1	Historical context	291
24.2	Awareness	291

	24.3	Minimum competence level	292
	24.4	*Vade mecum*	294
	24.5	Trouble with *some* standards	296
	24.6	In practice	297
	24.7	What could be argument(s) against ...	298
	24.8	Practice, practice, practice...	299
	24.9	The last word	302
	24.10	References	302
	24.11	Further reading (a first selection)	304

25 Representative sampling and society ... 307

	25.1	Sampling: from the point of view of buyers, consumers, citizens	307
	25.2	The way forward: some proposals	311
	25.3	Beyond traditional application fields	313
	25.4	Conclusions	316
	25.5	References	317

26 Epilogue: what's next? ... 319

Foreword

This book is based on the series of Sampling Columns published in *Spectroscopy Europe/Asia* between 2014 and 2019. However, there has been a great deal of updating and editing of what originally appeared for this book. While the subject matter is closely related to that in the columns, "A newcomer's introduction to representative sampling", the way this book goes about its task is greatly influenced by the author's extensive experience teaching this topic *live* to a wide range of audiences over two decades. The book is being published 21 years since the author met the creator of the Theory of Sampling (TOS), Pierre Gy, for the first time, 20 December 1998.

This book is *not* a textbook proper, and neither is it a mere lecture compendium—it is a passionate INTRODUCTION in the form of an augmented rendition of the way a typical three-day course on "Theory and Practice of Representative Sampling" manoeuvres through what undoubtedly will be a bewildering and complex topic at first sight. Teaching the TOS is a fascinating challenge—every time. No two sessions are alike; inspired teaching is very much like *dancing* with an audience. It is my hope that this relationship shines through when the reader is working through the book's chapters. There are several ways to try to bring clarity to the TOS (and there are several parallel introductions available; the third edition of one pre-eminent, high-level textbook was published in 2019). However, **the present book is the simplest and easiest way to acquire the necessary and sufficient overview of the TOS for all newcomers to this topic!** If, upon completion, the reader does not agree with this self-assured statement, please feel mandated to submit suggestions for improvement!

Credits
Original collaborators are acknowledged in the individual *Spectroscopy Europe* columns that served as a basis for many of the chapters in the book (https://www.spectroscopyeurope.com/sampling, 👆 bit.ly/tos20-2)

More credit
Also, much praise and thanks to Brad Swarbrick, Oscar Dominguez, Rodolfo Romanach and Anne Jodon Cole for critical proof reading. The author and the Publisher are grateful for their willingness to do this essential task which has improved the book significantly.

I want to thank many individuals and institutions for contributing to developing my own TOS overview: my students over the last two decades, academic and industrial colleagues, members of government and regulatory authorities, co-participants to workshops, symposia, conferences a.o. Many will be named within the covers of this book, but here I specifically want to express my overwhelming gratitude to Publisher, Editor-in-Chief and brother-in-arms re. all matters related to how to get the TOS out to *new audiences*, Mr Ian Michael. Ian in many ways is the opposite to this author: where for me it is *all* about the creative teaching and writing process (as in *ALL*), it is sadly not also about the necessary discipline and ability to keep track of a whirlwind of notes, chapter versions, illustrations and much less about keeping order in one's own files etc.: "Where is the fun in that?". However, without Ian's total professionalism regarding all these equally important matters, there would assuredly *never* have been any book. **Thank you so much Ian** (this goes for the previous book we produced together too)!

This book has benefited significantly from having a major sponsor along for the ride, all the way. The company **HERZOG** is equally interested in contributing to raising the awareness of and expertise in the TOS in all relevant industrial sectors. HERZOG has been a much-valued support for the long haul behind this book, both providing significant financial sponsorship, as well as, importantly, by allowing Mr Martin Lischka to create very many professional illustrations. I thank CEO, Mr Jan Herzog, and Martin Lischka profusely for their contributions.

The greatest thank you goes to my wife, Anne Jodon Cole, for her loving care and understanding, patience and insight concerning this work of mine—which for an outside observer easily could run the risk of being dismissed as mere *workaholism*. This would be a total misunderstanding however, concerning only a quantitative aspect. My professional career has been a life-long didactic

Foreword

mission, without which there would be very little meaning to it all.... Allowing this to be a very large part of our life together is a great gift she has given me, for which I am forever grateful. And there is so much else I am grateful for: Ellington to you!.

Kim H. Esbensen

Copenhagen, December 2019

References throughout the book

We have made strenuous attempts to add useful References for further reading wherever appropriate. Many of these are freely available, whether Open Access journal papers or available in other ways. To avoid the reader having to type a long and often confusing URL into their browser, we have used the link-shortening service bit.ly to produce easier to type URLs. These are identified by the 👆 symbol. They (almost always) have the format bit.ly/tos1-1. The characters after "bit.ly/tos" refer to the chapter and reference number. In a number of cases, a reference will be used in more than chapter, in which case the first shortened version is used. Since most browsers add the http:// automatically, all the reader needs to do is type "bit.ly/tos1-1", for example, into their browser.

We have also created a web page where you can find all the References from this book in one convenient place. Just open this when reading the printed book and enjoy easy access to all the References with one click: https://www.impopen.com/tos-references.

About the author

Kim H. Esbensen, PhD, Dr (hon), research professor in Geoscience Data Analysis and Sampling at GEUS, the National Geological Surveys of Denmark and Greenland (2010–2015), chemometrics & sampling professor at Aalborg University, Denmark (2001–2015), professor (Process Analytical Technologies) at Telemark Institute of Technology, Norway (1990–2000 and 2010–2015) and professeur associé, Université du Québec à Chicoutimi (2013–2016). From 2015 he phased out a more than 35-year academic career for a quest as an independent consultant: www.kheconsult.com. However, as he could not terminate his love for teaching, he is still on a roll as an international visiting-, guest- and affiliate professor around the world.

A geologist/geochemist/data analyst by training, he has been working for three decades at the forefront of chemometrics, but since 2000 he has devoted most of his scientific R&D to the theme of representative sampling of heterogeneous materials, processes and systems (Theory of Sampling, TOS), and PAT (Process Analytical Technology). He is a member of several scientific societies, has published over 250 peer-reviewed papers and is the author of a widely used textbook, *Multivariate Data Analysis* (35,000 copies), published in its 6th edition in 2018. He was the originator and chairman of the taskforce behind the world's first horizontal (matrix-independent) sampling standard DS 3077 (2013). He is editor of the science magazine *TOS forum* (https://www.impopen.com/tos-forum) and for the Sampling Column in *Spectroscopy Europe/Asia* (https://www.spectroscopyeurope.com/sampling).

Esbensen is fond of the right kind of friends and dogs, swinging jazz, fine cuisine, good wine, contemporary art and classical music. His has been collecting science fiction novels for more decades than he is comfortable contemplating, still, as ever, it's all in the future…

https://kheconsult.com

International Pierre Gy Sampling Association (IPGSA): https://intsamp.org

YOUR PARTNER FOR
- AUTOMATIC SAMPLE PREPARATION -

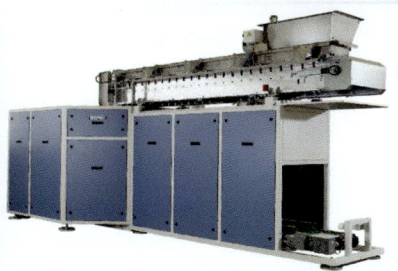

HERZOG
Get your sampling right

YOUR PARTNER FOR *Get your sampling right*
- AUTOMATIC SAMPLE PREPARATION

HERZOG is the world's leading manufacturer of equipment and automated solutions for quality control in the minerals and materials extractive and recycling industry. From primary sampling through sample preparation and analysis - HERZOG provides all components, tools and accessories for automated process monitoring. As a leading supplier of automated sample preparation laboratories HERZOG Maschinenfabrik GmbH & Co. KG provides equipment based on 60 years of experience.

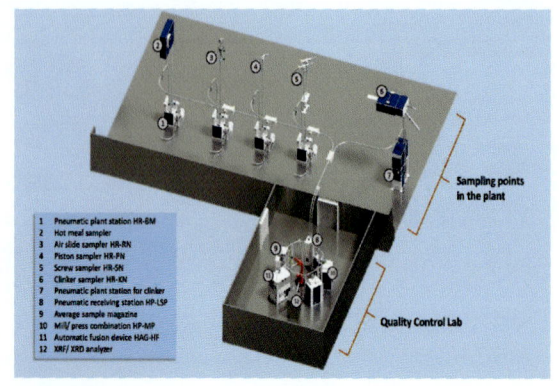

Fully automatic sample preparation solution covering primary sampling, sample transport and presentation of the final aliquot to the analyzer (cement application illustration)

1. Pneumatic plant station HR-BM
2. Hot meal sampler
3. Air slide sampler HR-BN
4. Piston sampler HR-PN
5. Screw sampler HR-SN
6. Clinker sampler HR-KN
7. Pneumatic plant station for clinker
8. Pneumatic receiving station HP-LSP
9. Average sample magazine
10. Mill/ press combination HP-MP
11. Automatic fusion device HAG-HF
12. XRF/ XRD analyzer

Industrial sampling & preparation solutions

Rotary Sample Splitters (RSS) are used extensively in automated laboratories. After primary sampling, size-reduction has to be undertaken in such a way that the final aliquot presented is fit-for-purpose representative of the original lot, or the primary sample delivered by the customer.

Often sub-lot sizes up to several hundred kilograms must be mass-reduced and split representatively, secondary and tertiary composite samples have to be created from which only a few grams will be the basis for analysis. This places the outmost demands to representativity in all sub-sampling, processing and aliquot preparation steps.

Splitters are available in various sizes and configurations and have to be chosen with respect to the comprehensive knowledge given by the Theory of Sampling, TOS.

Typical sampling variances have to be estimated using the RSV measure. RSV = Relative Sampling Variability, a measure for quantifying the effective, total sampling + analysis variability as expressed by a Replication Experiment (RE), see DS3077 (2013)[1].

Fully automatic rotary splitter for fine-powered materials with maximum particle size of 500 µm. An RSV of 5 % has been verified by replication experiments.

1. DS3077 (2013) "Representative Sampling – Horizontal Standard" www.ds.dk

Sampling equipment	Splitting equipment	Knowledge transfer	Software control
Primary sampler (Moving & Stationary lots)	Rotary splitting	Seminars	Sample taking (PrepMaster)
Secondary & tertiary cutters	Linear splitting	Academic cooperations	
Special sampler designs	Slurry cutter	Sampling system validation support	Quality monitoring (PrepMaster Analytics)
Sample transportation solutions	Special splitter design	Validation of measurement systems	
Mixer/Blender			

There is a wide commercial potential for automated laboratory solutions in PGM recycling and in many other materials and commodity sectors, e.g. mining, food, feed, pharma, secondary raw materials, recyclates in general, agricultural products and seeds.

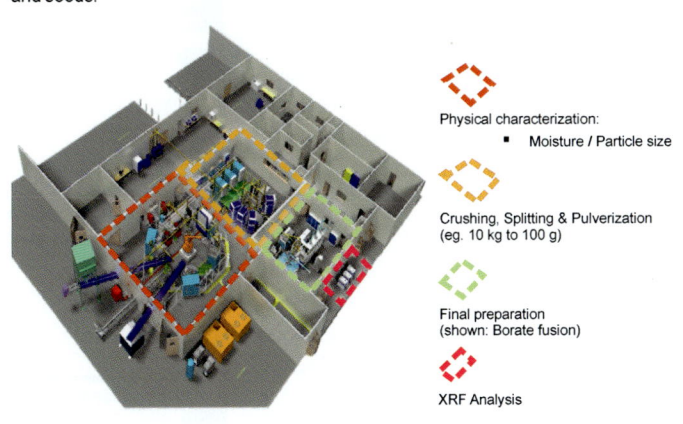

Physical characterization:
- Moisture / Particle size

Crushing, Splitting & Pulverization (eg. 10 kg to 100 g)

Final preparation (shown: Borate fusion)

XRF Analysis

Typical mine site automated laboratory for exploration samples, semi-finished products and final products. Such automation covering the whole sample preparation workflow from "sample delivered to the laboratory to final aliquot presented to the analysers of choices" are successfully established for multiple industrial applications.

Special sampling equipment

In close cooperation with our customers HERZOG develops special sampling equipment for various non-standard application fields. HERZOG is a responsible partner from the early steps in design of a new sampling protocol, and the necessary equipment - to installation and verification of new prototypes. Subsequent crushing and splitting equipment can be added and implemented as needed. All equipment design follows the guidelines given by TOS with the aim of offering the lowest Relative Sampling Variance (RSV) and for realizing fit-for-purpose solutions tailored for each customer's specific needs.

Teaching and research

Representative sampling is a critical success factor for achieving the highest quality analytical accuracy and precision needed in process control. Mastering primary sampling is especially important for the automated analytical laboratory. All steps in the complete lot-to-aliquot pathway are compliant with DS3077 (2013) standard for representative sampling.

In collaboration with KHE Consulting, Copenhagen, HERZOG is offering consulting, audits, courses and seminars (general, or in-house) to our customers on the principles and practice of representative sampling and sub-sampling. A shared TOS competence between OEM HERZOG and customers guard against the many pitfalls impacting the total sample preparation error in the automated laboratory.

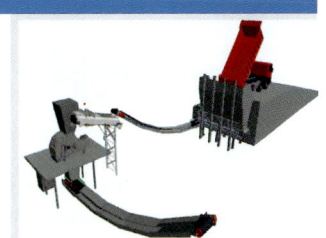

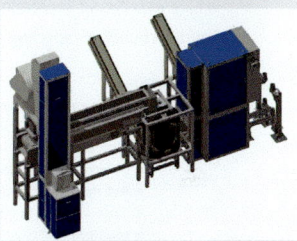

Primary sampling
Secondary sampling
Crushing
Representative splitting
Grinding
Representative splitting
Fine grinding
Final preparation
Analysis
Data evaluation

HERZOG
Get your sampling right

PrepMaster Analyitcs

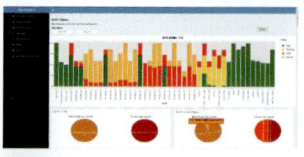

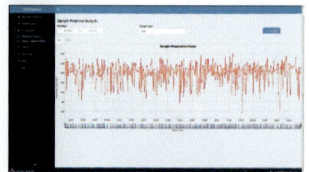

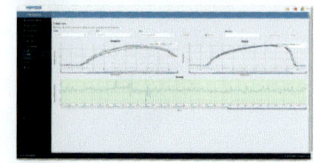

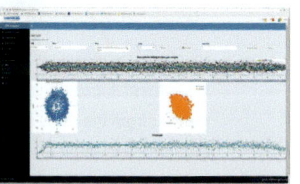

PrepMaster Analytics (PM) is HERZOG's proprietary software for ingestion, storage and evaluation of analytical and sensory performance data as well as for on-line processing monitoring. PM Analytics is specifically tailored to the requirements of automated laboratory systems and provides an easy and holistic view on current Key Performance Indicators of your processes (KPI). Due to a flexible software design PM Analytics can be readily embedded into the software environment of each customer.

PM Analytics uses modern big data frameworks for fast computing of large quantities of individual data points from a diversified array of sources. PM Analytics comes with a fully web-based approach, responsive design, clean and polished dashboards as well as a bunch of reporting and statistical tools.

At a glance

- Display all information's on web-based dash boards
- Pseudo-element calculation
- Data audit
- Linking analytical data to system KPI's
- Automated customized reports (Email, Mobil phone, …)
- Open interface to other LIMS systems and high level Scada systems
- Maintenance scheduler & Supervision by machine downtimes
- Tool condition monitoring (TCM):
  - Sensory data
  - Video surveillance assisting failure identification
  - Cost optimizing consumables
- System analysis:
  - Sample processing times
  - Message statuses
  - Error message archive
  - Laboratory performance analysis

CONTACT

GERMANY
HERZOG Maschinenfabrik GmbH & Co. KG

Auf dem Gehren 1
49086 Osnabrück
Germany

+49 541 9332-0
+49 541 9332-33
info(at)herzog-maschinenfabrik.de
www.herzog-maschinenfabrik.de

JAPAN
HERZOG Japan Co., Ltd.

3-7, Komagome 2-chome
Toshima-ku
Tokio 170-0003
Japan

+81 3 5907 1771
+81 3 5907 1770
info(at)herzog.co.jp
www.herzog.co.jp

USA
HERZOG Automation Corp.

16600 Sprague Road, Suite 400
Cleveland, Ohio 44130
USA

+1 440 891 9777
+1 440 891 9778
info(at)herzogautomation.com
www.herzogautomation.com

CHINA
HERZOG (Shanghai) Automation Equipment Co.,Ltd Section A2,2/F, Building 6, No.473, West Fute 1st Road, Waigaoqiao F.T.Z, Shanghai, 200131, P.R. China

+86 21 50375915
+86 21 50375713
MP: +86 15 80 07 50 53 3
xc.zeng(at)herzog-automation.com.cn
www.herzog-automation.com.cn

How to get your sampling right on

Linked in

YouTube

1 Theory of Sampling (TOS)—the missing link before analysis

What is the meaning of analysing a non-representative sample? None—it is pointless!

Undoubtedly everybody agrees with the answer to this provocative question. If a sample cannot be documented to be representative of the lot/target material from where it was taken, it is a waste of time, effort and money to analyse it! Without a minimum of sampling knowledge and experience, there is absolutely no guarantee that a "repeated sampling operation" will result in a "duplicate sample" with the same analytical result. This will create chaos and despair for the entity trying to make an informed decision based on analytical results. This scenario sums up precisely the background and rationale for this book.

Why is the above true? The answer is a universal attribute of all types of materials met with in science, technology and industry: *heterogeneity*, of which much will be said in this book. While there is plenty of "sampling" going on in these realms, much of it unfortunately cannot be characterised as being anywhere near representative. Indeed, this kind of sampling is rather often oriented towards easy, not-too-expensive activities, preferably carried out with the least effort. But this approach cannot result in the desired objective, representative samples, as shall be shown forcefully in this book.

However, sampling *can* certainly be both practical and representative, and not demand unreasonable efforts either. All it

AA

TOS: Theory of Sampling
NIR: Near Infrared
SE: Sampling Errors
CSE and ISE: Correct and Incorrect Sampling Errors
SUO: Sampling Unit Operation
RE: Replication Experiment
MU: Measurement Uncertainty
WCSB: World Conference on Sampling and Blending
IPGSA: International Pierre Gy Sampling Association
PAT: Process Analytical Technology
TSE: Total Sampling Error
TAE: Total Analytical Error
a_L: True, average concentration of the analyte of interest in the lot

A unit operation is a process where materials are input, a function occurs and materials are output. A unit operation is any part of potentially multiple-step processes which can be considered to have a single function. Examples of unit operations in chemical engineering include: separation, purification, mixing, reaction processes, heat exchange and power generation processes. The Theory of Sampling has adopted this concept in order to be able to treat four TOS processes with maximum conceptual clarity: composite sampling, crushing, mixing and mass reduction—all of which will be explained and expanded upon throughout this book.

needs is to follow the principles and unit operations outlined by the Theory of Sampling (TOS).

The ever-increasing precision with which quantitative analytical modalities are able to perform is often based on an inversely decreasing analytical mass (test portion). The trade-off is that the higher the analytical precision desired, the smaller the volume to be characterised. There may be attempts to counteract this by efforts to increase the "effective sample presentation volume". For example, for quantitative near infrared (NIR) spectroscopy the static Petri dish may be supplanted by a rotating dish, then by the roll bottle, to be replaced by.... Though the degree to which this same trend is manifested in many different analytical methods may vary, the common issue is that the test portion continues to be but a very small fraction of the original target material (termed "lot" in the TOS).

More importantly, in the broader perspective, precision cannot be traded for accuracy. Irrespective of increased analytical precision (which is a matter solely related to the analytical method and equipment), the accuracy of an analytical result is ultimately supposed to be an attribute related to the analyte content determination of the *complete lot* from which the test portion was derived. In fact, this relevant accuracy is a matter almost solely related to the sampling process and, as will be shown, the analytical accuracy often all but vanishes in this context. Barring trivial micro- and meso-scale laboratory examples from which no generalisation can be made, it is readily acknowledged that to come from a typical lot in science, technology or industry to the analytical aliquot, sampling must have been involved. It is not always equally readily understood that the sampling rates involved are typically orders of magnitude $1:10^6$, $1:10^9$ or $1:10^{12}$ (Figure 1.1). Sampling spanning such a number of orders of magnitude (mass:mass or volume:volume) to produce the analytical aliquot is far from a simple operation, because all materials

Figure 1.1. Where representativity starts, at the primary sampling stage—long before analysis. This generic illustration is a placeholder for all the world's infinitely different types of materials and lots. It also illustrates the crucial point that heterogeneity need not necessarily be visible to the naked eye.

are inherently heterogeneous (at some scale or other, which will be explained in detail). Sampling of heterogeneous materials is most emphatically not a simple bulk materials handling issue! Heterogeneity is both the common denominator of all types of materials (of all lots), as well as the arch-enemy of all sampling efforts, Figure 1.2.

There is not enough appreciation, understanding and appropriate skills around the critical role of representative sampling. Here is the case in point. When confronting a lot (of any size, the issues below are scale-invariant) with a sampling purpose (with a sampling implement in hand—scoop, spear, whatever—addressing either a stationary or a moving lot), the following question would seem to be fully justified: "How big a sample is needed, in order for it to be representative?" Surprisingly, upon reflection guided by the TOS, this is the wrong question, at the wrong time, in the wrong place! For lots of any size, the sampling issue, from almost all points of view, can easily become overwhelming and simplified approaches are invariably looked for—the "how big a

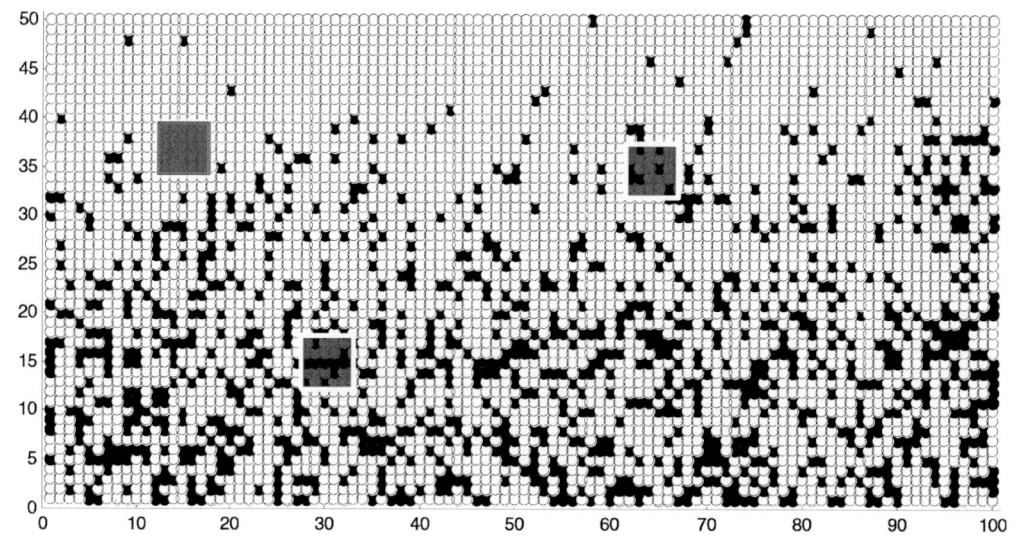

Figure 1.2. The arch-enemy of representative sampling: *heterogeneity*. A "grab sample" (a single event sampling) can never be representative of a heterogeneous lot, a theme to be treated exhaustively in this book. The fact that the three grab samples shown will have radically different analytical results (black: analyte) is not a fault due to the analytical laboratory, but is a powerful illustration of what is called the Fundamental Sampling Error (FSE), to be explained in full throughout this book. However, consider these three extractions as "increments" and aggregating them—to a composite sample. More increments will surely be needed, but the principle of *composite sampling* is already well indicated.

sample must I take in order for it to be representative..." is by far the reaction most often met with. Sadly, it is utterly wrong.

The wrong question: because representativity is not related to sample size, but to the sampling process—ascertaining whether a particular sample is representative, or not, *cannot* be resolved by characterisation of the sample itself. This may be news to many.

The wrong time: because this issue should have been thought through long before the actual sampling commences. This issue cannot be solved at the same time as one is preoccupied with fixing the sample volume, i.e. sample size or mass—which in any way is not the driver for representativity (another surprise?).

The wrong place: unless one has not already started learning a certain minimum of proper sampling principles, it is most likely that the focus is on taking just one sample, as this is indeed the easiest—a procedure termed "grab sampling". However, the most important tenet of the TOS is that grab sampling is always wrong—and that only composite sampling (multi-increment sampling) is able to lead to and to guarantee representativeness.

This book deals extensively with all these issues. In composite sampling, a *sufficient* set of individual increments covering the entire volume of the lot is essential; determining the numerical issue of "sufficient" is part of the definition of representativity and is, in fact, the correct answer to the question of "how big a sample". All shall be explained below.

1.1 A framework for representative sampling

This book deals with the universal principles of representative sampling (as opposed to mere mass reduction, which is concerned with sample masses only), presenting to the reader a comprehensive, necessary and sufficient body of knowledge with which to be able to perform, and document, representative sampling at all scales, for all types of materials (lots). Some of the TOS may be known to the reader, at least partially, but new insight will be presented in the form of a coherent, universal framework, and there will be some surprises as well. First up is an overview of the elements in such a framework in the briefest possible format:

- Why sampling? (why materials handling is not sampling)
- The TOS—fundamental definitions
- Sampling terminology (the tower of Babel)
- The sampling target, the lot and lot dimensionality (0-, 1-, 2-, 3-dimensional lots)
- Heterogeneity—the arch-enemy of sampling
- Representativity—a formal definition

- Representativity is solely a characteristic of the sampling process
- Representative sampling is always a multi-stage process
- Introduction to the concept of sampling errors (SE)
- Correct and incorrect sampling errors (CSE, ISE)
- Process sampling errors
- Sampling Unit Operations (SUOs)
- Special focus on mass reduction (one of four SUOs)
- Heterogeneity characterisation
- The Replication Experiment (RE) for stationary lots
- Variographic analysis (for dynamic lots, i.e. moving lots)
- Practical process sampling (sampling of moving 1-D lots)
- Sampling in the 2-D plane (what is special about 2-D?)
- Four Quality Criteria to ensure representative sampling
- The TOS vs Measurement Uncertainty (MU)—a call for integration
- The analytical bias is constant—but the sampling bias is **not**
- International standards, guidelines, norm-giving documents
- DS 3077 (2013) Horizontal—A unified standard for representative sampling
- Pierre Gy—the founding father of the TOS and a monumental scientific *oeuvre*
- World Conferences on Sampling and Blending (WCSB)
- Sampling Hall of Fame/Shame (instructive case histories from which to learn)
- TOS literature (further information sources)
- International Pierre Gy Sampling Association (IPGSA)
- *TOS forum*—the scientific magazine of the IPGSA[1]
- *Spectroscopy Europe/Asia* "Sampling Columns"

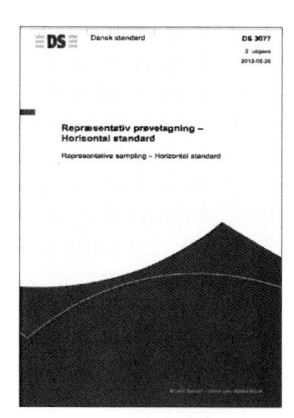

Figure 1.3. The world's first matrix-independent ("horizontal") standard on representative sampling provides the most concise, comprehensive introduction to the TOS available, powerfully augmented by the additional, selected references in this chapter.

TOS forum:
www.impopen.com/tosf
bit.ly/tos1-7

Spectroscopy Europe:
www.spectroscopyeurope.com/sampling
bit.ly/tos1-8

1.2 What comes before analysis—the TOS!

The following scene-setting is valid for pretty much all types of final analysis, spectroscopic or otherwise, carried out either in the laboratory or on location, at the manufacturing or production line, at the conveyor belt or at/in the pipe line (Process Analytical Technology, PAT).

In a quite specific sense, it is *all* about what comes *before* analysis, in the sense of the provocative question stated in the introduction: what is the meaning of analysing a sample if it cannot be documented to be representative of the lot? So how does one achieve and document that a specific sample is indeed representative of the target material?

Here comes the first surprise. It is not possible to document that a specific sample is representative by any known method, approach or activity directed at the sample itself—be this analytical, data analytical, statistical or otherwise. It is not possible to discern the status of a given sample from any type of inspection of the sample itself. All specific samples, when *observed in isolation*, only allow one characterisation—they constitute a very small, mass-reduced fraction of the lot (representative or not). But mass reduction in and of itself has **nothing** to do with representativity. It is only the specific *sampling process* with which a sample was extracted that can be designated as representative, or not, according to certain criteria which will be presented in this book. After a sample has been successfully extracted (by a representative sampling process), it will have the weight necessary to achieve this goal—not the other way around, see above regarding "the wrong issue, at the wrong time and place".

In general, there must always be *sampling* from a lot; there must in all likelihood also be sub-sampling (one or more mass-reduction steps) in order to furnish the ultimate goal: the analytical aliquot. It is fatal to ignore these pre-analysis steps and only focus on the final analytical activity. As will become abundantly clear, the potential sum-total of the sampling and sub-sampling errors typically dominate in the total uncertainty budget compared to the analysis error effects alone. It is not unusual for this total sampling error to surpass the analytical uncertainty by factors of 10–20–50 or more, depending on the degree of heterogeneity of the material sampled and on the degree to which the sampling process has been purged of its own complement of possible sampling errors. How to characterise and quantify heterogeneity and, even more important, how to *counteract* heterogeneity in the sampling process by applying the relevant TOS principles and unit operations are all discussed below.

Designing a sampling process is a futile undertaking if not related to the material heterogeneity encountered. The TOS includes specific guidelines as to how to estimate lot heterogeneity, both for stationary as well as for dynamic (moving) lots. At the same time as this is accomplished, one will actually be able to optimise the specific sampling process involved (look for the "Replication experiment" and "Variographic characterisation" sections below).

> **"A certain TOS minimum…"**
> Samplers must be conversant with the basic principles of the TOS, as must their supervisors and management… as must OEM companies and their sales forces. It is manifestly not enough to declare that "this equipment provides representative samples". There are a lot of easy promises out there… Clients and buyers should demand proof (TOS principles) from sellers and companies that promise the world "if only you buy this equipment from us…"

A certain minimum competence regarding the TOS must be acquired by any professional sampling operator or other personnel procuring samples, sub-samples and analytical aliquots, otherwise the prospects of documentable representativeness will be forfeited. As it turns out, the exact same principles govern sampling at all scales (primary, secondary…), which will make internalisation of the TOS much easier than perhaps might be thought initially. Plenty of relevant literature exists, at all levels of interest, that will allow anybody interested to acquire this skill, with no exceptions. Below are listed a first set of key introductory references to get the reader started[1–7] in which can also be found a plethora of background TOS literature.

1.3 What this book promises...

The introduction presented in this book is based on 20 years of experience teaching the TOS at several levels in science, technology and industry: in academia (BSc, MSc, PhD levels), for companies and corporations, identically for the same educational levels, for all entities charged with the responsibility for sampling, or who work in the field or in plants, including process technicians—and finally for management (across all levels, top, middle and operational management). This book will focus on all aspects of sampling theory (TOS), sampling practice and the economic consequences of not taking sampling seriously. The curriculum in this book is equally directed at all individuals, agents, supervisors and technicians responsible for optimising plants and processes. The message in this book also applies to many other areas where *someone* must collect samples that ultimately furnish the foundation for making important decisions, but where this *someone* only rarely has been properly trained with respect to this critical task. Thus, this book is primarily for the uninitiated beginner to the world of conscientious, responsible and defensible sampling—for all those motivated and ready to put in the surprisingly modest amount of effort needed to improve their appreciation, overview, competence and skill sets with which to become professional samplers. Welcome to the world of the TOS!

The hallmark of this book is its unique didactic style, specifically designed to convey a comprehensive overview of what has traditionally been considered as a very complex matter. The emphasis is on simplifying overarching concepts, principles, sampling error management rules, as well as the practical procedures and operations making up the elements of the TOS, and how these interact: see the back cover of the book for an initial idea of how all this is interrelated.

This book serves as a basis for a self-study introduction to the TOS. As such it is not a textbook in the classical sense, but a book version of the current form of the introductory course as developed by the author over 20 years. Thus, the reader will likely recognise a few iterations along the way: select definitions, explanations repeated exactly where experience has shown that they would be most welcome amidst a steady onslaught of new information, interim summations, key scientific points... even a few repeated illustrations, all where they would appear in an oral course format. This is of no matter, however, in view of the ambitious didactic objective of the book: this is how the complicated curriculum is best presented, aimed at delivering the professional overview promised to all newcomers to the TOS. In fact, this book will present an overview that is also novel to quite a number of the current sampling community.

How to read this book

The reader is encouraged to read the first two chapters, followed by Chapters 22–24, and only then continuing from Chapter 3. In this fashion the reader is guaranteed optimal conditions with which to achieve all the promised learning goals.

The reader is offered a plethora of references, indeed some will appear several times, associated with different chapters. A major design principle of this book is to offer as many as possible illustrating and/or higher level additions complementing the text. Many of these are freely available to download and read, and can be easily accessed from https://www.impopen.com/tos-references.

The reader is recommended to consult Reference 1 to complement this opening chapter with more comprehensive explanations; Reference 1 is available on-line.

1.4 References

1. K.H. Esbensen and C. Wagner, "Why we need the Theory of Sampling", *Analytical Scientist* (2014). https://kheconsult.com/wp-content/uploads/2017/11/WHYweneedTOS-TAS-short.pdf, 👆 bit.ly/tos1-9
2. DS 3077, *Representative sampling—Horizontal Standard*. Danish Standards (2013). http://www.ds.dk, 👆 bit.ly/tos1-2
3. K.H. Esbensen and L.P. Julius, "Representative sampling, data quality, validation—a necessary trinity in chemometrics", in *Comprehensive Chemometrics*, Ed by S. Brown, R. Tauler and B. Walczak. Elsevier, Vol. 4, pp. 1–20 (2009). https://doi.org/10.1016/B978-044452701-1.00088-0, 👆 bit.ly/tos1-3 (N.B. new revised edition 2019)
4. K.H. Esbensen, C. Paoletti and N. Theix (Eds), "Special Guest Editor Section (SGE): sampling for food and feed materials", *J. AOAC Int.* **98(2)**, 249–320 (2015). http://ingentaconnect.com/content/aoac/jaoac/2015/00000098/00000002, 👆 bit.ly/tos1-4
5. K.H. Esbensen, "Materials properties: heterogeneity and appropriate sampling modes", *J. AOAC Int.* **98**, 269–274 (2015). https://doi.org/10.5740/jaoacint.14-234, 👆 bit.ly/tos1-5
6. K.H. Esbensen and C. Wagner, "Theory of Sampling (TOS) versus Measurement Uncertainty (MU)—a call for integration", *Trends Anal. Chem. (TrAC)* **57**, 93–106 (2014). https://doi.org/10.1016/j.trac.2014.02.007, 👆 bit.ly/tos1-6
7. https://www.impopen.com/tosf, 👆 bit.ly/tos1-7

2 Theory of Sampling (TOS)—fundamental definitions and concepts

The Chinese sage Confucius is reported to have stated: "Speak precisely—and wars can be avoided".

This is powerful insight, addressing what is needed for effective communication. The meaning is clear: one must at all costs avoid uncertainty, imprecision, vagueness in oral and written communication. Exactly the same holds for those who want to communicate in science, technology and industry, especially concerning a topic that traditionally has been considered "difficult"—the TOS. It is crucially important to be able to speak with the utmost precision.

This chapter introduces the most important fundamental definitions and principles of the TOS without which no rational understanding and appreciation can be established.

Samples are extracted for various reasons, using different sampling procedures in a wide range of application fields addressing a bewildering array of different material types. One would think that many potentially different sampling procedures would be needed. However, the main purpose of sampling is the same—to be able to extract a reliable small mass of the target lot that is to be characterised (analysed), i.e. to obtain a sample, which accurately and precisely *represents* the lot (see definitions below). Sample representativity is, therefore, the first, indeed the only, criterion that must be honoured in order to be able to draw valid conclusions about the characteristics of an original lot. Non-representative samples ("specimens" in the TOS) will result in an

Photo: Wikipedia

A*A*
ISE: Incorrect Sampling Errors
SUO: Sampling Unit Operation
GP: Governing Principles (of the TOS)
CH: Compositional Heterogeneity
DH: Distributional Heterogeneity
FSP: Fundamental Sampling Principle

unavoidable risk of erroneous decisions and conclusions without any possibility of knowing to what degree this is the case.

The TOS defines sampling as a *multi-stage* process, allowing a complex task to be broken down into its element stages and individual, or any required combination of, SUOs to be applied to cover all situations. The TOS's sampling unit operations will be described in detail, with a major focus on the *sampling process* and not the sample itself. Once extracted, there is no possibility of evaluating whether a *specific* "sample" is representative of a target lot, or not. It is the sampling process that is the **only** guarantee for obtaining a representative sample. Disobeying or compromising the TOS's principles will unavoidably lead to non-representative sampling procedures. These distinctions will be more fully defined and discussed below.

This chapter presents a minimum set of terms and definitions in the TOS, without which it is not possible to fully understand, nor apply, the TOS in a meaningful way. It is convenient to start by defining "lot dimensionality".

2.1 Lot dimensionality

The special case of a zero-dimensional (0-D) lot refers to a lot that can be effectively mixed, moved and sampled throughout with ease and complete correctness. Usually these are small lots, which can easily be manipulated at the laboratory workbench. A full definition of the 0-D lot is given in the terminology section of DS3077.[1]

Lot dimensionality is characterised by specifying the number of effective dimensions that need to be *covered* by the sampling process. This approach allows definition of one-, two- and three-dimensional (1-D, 2-D and 3-D) lots as well as the "zero-dimensional" (0-D) lot. Figure 2.1 compares the TOS's four cases of lot dimensionality.

The concept of lot dimensionality becomes clear, for example, when considering an elongated material stream, such as the case of dynamically moving material on conveyor belts; or a flux of matter ducted in a pipeline (Figure 2.2). Such lots can loosely be described as one-dimensional, since one dimension of the physical geometrical aspect dominates (the conveyor belt or pipeline transportation direction). However, according to the TOS, it is essential to consider how the specific sampling method applied *interacts* with the effective number of dimensions during the sampling process. Employing grab sampling (extracting a single increment as a "sample") on such an elongated material stream, which is a widely-applied but fundamentally flawed extraction method, would in reality make this a 3-D lot *not* a 1-D lot, since grab samples are most likely only taken from the top surface part of the moving material flux, and so are far from *covering* both the transverse lot dimensions fully, i.e. the full width and thickness. This geometrical "covering" aspect is a fundamental issue for the TOS.

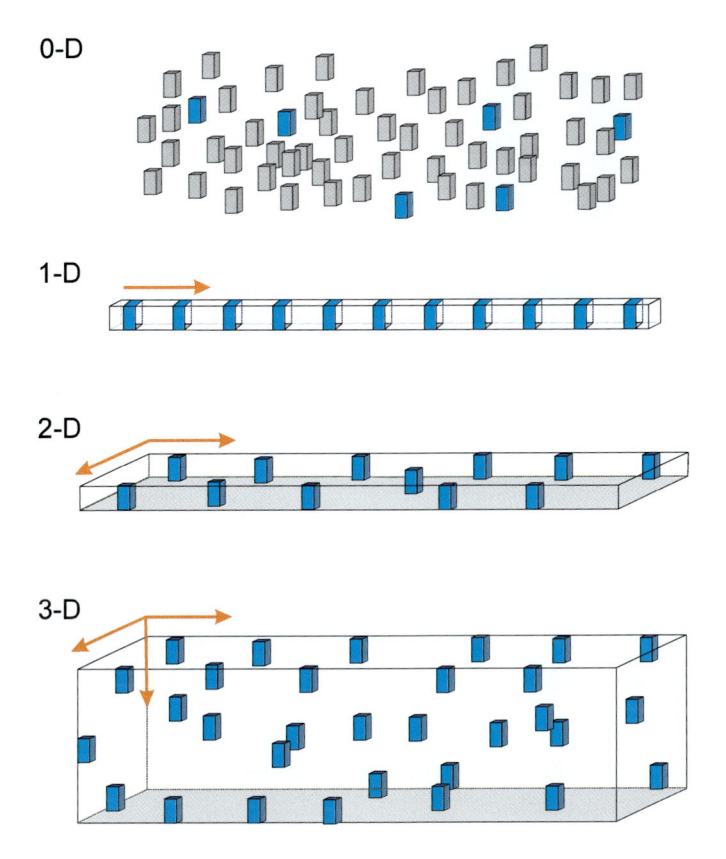

Figure 2.1. Lot dimensionalities: 0-D, 1-D, 2-D and 3-D lots. Potential increments are marked in blue. Note how the extracted increment can be made to cover the transverse dimensions only for 1-D and 2-D lots but not for 3-D lots. The special 0-D lot is defined in full in the text. Note that an increment in the 3-D lot case ideally *should* be a sphere, but as this is a physical impossibility, 3-D increments are very nearly always extracted in the form of a cylinder, often in the form of a drill core. This is also often the case for 2-D lots, but see Figure 2.2 for an alternative increment extraction.

Figure 2.2. Lot dimensionality in practice. Upper left: 1-D lot; right: 2-D lot; lower left: 3-D lot. The TOS outlines that the principles behind representative sampling are scale-invariant so these examples also represent much smaller and/or much larger lots. Note how a conventional spade is used for increment extraction from the 2-D lot, cf. Figure 2.1.

By contrast, a cross-stream cutter (a sampling device especially designed for elongated material fluxes, which will be discussed extensively later) will cover the entire depth and width of the stream, thereby fully reducing the sampling lot to one dimension, i.e. the longitudinal dimension of the material stream. See

Figure 8.4 for an example of this practical aspect of the definition of 1-D lot dimensionality.

According to the TOS, 1-D lots present an optimal sampling situation, preferred over 2-D and 3-D lots (e.g., industrial, geological or environmental strata, stacks, stockpiles, silos) which should, wherever possible, be *transformed* to comply with a 1-D sampling situation (more of this in Chapters 15 and 16). "Transformation" should not necessarily be viewed as a forced operation—in practice this is often possible by simply locating a situation where the lot already is *in transport*. Sometimes even original 0-D lots are also transformed into the desired 1-D configuration, because this offers unbeatable optimal sampling conditions even here. Lot dimensionality transformation constitutes one of the six governing principles (GPs) of the TOS, guiding the reader through this book.

Figure 8.4. Principal sketch of the archetype cross-stream cutter for 1-D lot sampling.

The reason for being so specific about lot dimensionality is the inherent heterogeneity of all naturally occurring materials, as well as raw materials and manufactured products, which makes sampling far from a trivial materials handling issue. Proper understanding of the complex heterogeneity phenomenon, its influence on sampling opportunities and, most importantly, of how heterogeneity can be *counteracted* in the sampling process requires a certain level of basic TOS knowledge. The purpose of this book is to gradually build up this knowledge. A first instalment of necessary definitions follows.

2.2 Sampling terminology—the tower of Babel

2.2.1 Lot

The lot is the complete entity of the original material being subject to sampling, e.g. a truck load, railroad car, process stream, ship's cargo, industrial batch etc. The lot (also termed the sampling target or decision unit) refers both to the physical, geometrical

Definitions in this chapter serve as initial information needed to get started, some will be augmented, and many more will be added as the introduction unfolds below.

form and size, as well as the characteristics of the material being subject to sampling, and specifically its heterogeneity.

2.2.2 Heterogeneity

Heterogeneity is the prime characterisation of all naturally occurring materials, but also of industrially manufactured or processed lots. Heterogeneity manifests itself at all scales related to sampling for literally all lot and material types. The only exception is *uniform materials*, which is such a rare example that no generalisation with respect to sampling can be made on this basis.

The TOS differentiates between two types of heterogeneity, one referring to the compositional differences between the individual "units" of the target material (compositional heterogeneity, CH) and one concerning the spatial distribution (distributional heterogeneity, DH) of the target material. The next chapter will deal extensively with these two types of hetergeneity. It is also convenient to distinguish a specific grain size heterogeneity (in itself part of CH), which sometimes manifests itself dramatically as a distinct lot DH characteristic as well, see for example Figure 5.5.

Correctly extracted material from a significantly heterogeneous lot can only come about due to an unbiased, representative sampling process. The term "sample" should always only be used in this qualified sense of "representative sample". If there is doubt as to this characteristic, the term "specimen" should be used instead. The critical term "representativity" should, of course, be very carefully defined and used with caution.

2.2.3 Specimen

A specimen is a "sample" that cannot be documented to be the result of a *bona fide* representative sampling process. It is not possible to ascertain the representativity status of any isolated

---

Uniform materials
Materials with a repeated (correct) sampling imprecision lower than 2%. Such materials do only occur very rarely in science, technology and industry (with the exceptions of gasses and infinitely diluted solutions etc.).

Sneak preview
Later, in several places, this book introduces the concept of a "heterogeneity contribution", termed h. h is a notion capturing the contribution to the whole lot heterogeneity incurred either by a fragment, or by an increment. The former is almost exclusively used in theoretical deliberations in the TOS, while the latter conceptualises the key aspect of practical sampling: quantitatively, the part of the whole lot heterogeneity that is carried by a particular sampled increment.

small part of a sampling target in-and-of itself. It is only the *sampling process*, which can be termed representative or not.

2.2.4 Correct sampling

The TOS uses this term to denote that specific efforts have been executed, which have resulted in successful elimination of the so-called "bias-generating errors", aka the Incorrect Sampling Errors (ISE). Incorrect sampling errors will be discussed in much more detail later. If not eliminated, ISE will lead to a troubling sampling bias.

2.2.5 Representativeness

Representativeness implies both correctness as well as a sufficiently small sampling reproducibility (sampling variance, or sampling imprecision).

2.2.6 Sampling bias

Systematic deviation between the average analytical sampling result and the true lot concentration, aka *accuracy*. Elimination of the sampling bias is the first obligation for any sampling process in order to be correct. This is accomplished by eliminating the ISE.

2.2.7 Reproducibility

Sampling variance, properly estimated only *after* removal of sampling bias. Sampling reproducibility is also known as sampling imprecision.

2.2.8 Increment

Correctly delineated and materialised practical sampling *unit* of the lot which, when combined with other increments, provides a multi-increment sample. This procedure is termed "composite sampling" in the TOS, with the result being a "composite sample".

A critical parameter for any composite sample is "Q", see immediately below.

2.2.9 Composite sample (Q)
Aggregation of several increments, the number of which is designated as Q. A composite sample represents "physical averaging", as opposed to arithmetic averaging of analytical results from individual increments. For composite sampling a necessary corollary is that the Q increment shall be deployed so as to "cover" the full lot volume as best as possible given Q, in order to uphold the Fundamental Sampling Principle (FSP), developed further below. There is a world of (sampling) difference between these two, perhaps at first sight, equal ways to arrive at an estimate of the average concentration of an analyte in a lot. Composite samples must always be subjected to thorough mixing immediately after extraction for reasons that will become clear below.

2.2.10 Sub-sample
The correctly mass-reduced part of sample (primary, secondary sub-sampling...). A sub-sample is a result from a *dissociative* (disaggregation) process; a composite sample is a result from an *integrative* process. It would be seriously misleading to call an increment a "sub-sample", although this distinction is (very) far from being observed in the scientific literature—a clear violation of Confucius' dictum.

2.2.11 Fragment
Fragment refers to the smallest separable unit of the material that is not affected by the sampling process itself (e.g. particles, grains etc.). By naming the smallest unit-of-interest a *fragment*, the TOS is also able to treat the situation in which the sampling process results in fragmentation of (some) of the original units. This

linguistic sleight-of-hand, that at first sight is perhaps surprising, in reality allows the TOS a totally unsuspected generalisation power.

2.2.12 Group

A number of spatially correlated fragments, which act as a coherent unit (increment) during sampling operations. While a group plays an important role in the *theory* of sampling, in practical sampling, the only group of interest is the actual *increment* being extracted, i.e. the material to be found in the sampling tool. The group size depends on the sampling tool (mass/volume) and the sampling process, as well as how the tool is implemented and operated.

2.2.13 Scale

The principles described by the TOS are *scale-invariant*, i.e. the same principles apply to all relevant scales and stages in the sampling pathway (lot, sample, sub-sample).

2.2.14 Zero-dimensional lot (0-D lot)

The 0-D lot is characterised by displaying no internal correlations between all potential increments, thus allowing for relatively easy practical sampling. A 0-D lot can be *manipulated* with ease— at least in principle—for example, by mixing or directed *in toto* splitting, but the work necessary may vary significantly as a function of the lot mass, M_L, and also because of other features, e.g. stickiness, irregular fragment forms.

For a full set of definitions, refer to the horizontal sampling standard *DS 3077*.[1] Additional explanations can be found in Reference 2 and will also be given where appropriate.

2.3 References

1. *Representative Sampling—Horizontal Standard*. Danish Standards DS 3077 (2013). www.ds.dk, bit.ly/tos1-2

Representative Sampling—Horizontal Standard. Danish Standards DS 3077 (2013). Preview: bit.ly/tos2-9

J. AOAC Int. **98(2)** (2015). https://www.ingentaconnect.com/content/aoac/jaoac/2015/00000098/00000002 bit.ly/tos2-11

Only one Theory of Sampling
There has been a debate on "Alternatives to Gy's Sampling Theory?" on LinkedIn. This is a good example of the critical need for precise speaking. We shall have ample occasion to return to this discussion later, but first after all the necessary basic concepts, definitions and principles have been properly introduced. For the specially interested, an example of incompetence with respect to the TOS has, sadly, been published in Wikipedia: https://en.wikipedia.org/wiki/Gy%27s_sampling_theory, bit.ly/tos2-1. This entry represents a gross misunderstanding of the TOS, and has been severely criticised and debunked by Esbensen & Lyman: https://doi.org/10.1255/tosf.11, bit.ly/tos2-1a

Public discussion of the TOS in theory, and especially in practice, can be found on various LinkedIn Groups, e.g.:

Theory of Sampling
👆 bit.ly/tos2-6

Society for Mining, Metallurgy & Exploration Group
👆 bit.ly/tos2-7

Industrial Sampling Systems Mineral Exploration Geoscience
👆 bit.ly/tos2-8

2. J. AOAC Int. **98(2)** (2015). https://www.ingentaconnect.com/content/aoac/jaoac/2015/00000098/00000002, 👆 bit.ly/tos2-5

3 Heterogeneity—the root of all evil (part 1)

What is common to all agents along the entire field-to-analysis pathway, e.g. process technicians and engineers, primary samplers, academic and industrial scientists, laboratory personnel, companies, organisations, regulatory bodies, and agencies as well as project leaders, project managers, quality managers, supervisors and directors? All are dependent on reliable analytical results to make decisions, but the analytical aliquots are only the final product of a multi-stage sampling process "from-lot-to-analysis". There is sampling—a lot of sampling and sub-sampling—before a *valid*, representative aliquot can be delivered to the analytical laboratory. All sampling has to deal with materials that are *heterogeneous* on one scale or another (or on all scales). It is vital to understand the characteristics of heterogeneous materials. One form or other of primary sampling is always necessary, which must *counteract* the effects of the sampling target heterogeneity. A total of six sampling errors arise because all sampling processes interact with the sampling target: two arise because of the material heterogeneity, and four additional sampling errors are produced by the sampling process itself. The latter, if not properly understood and eliminated, play a crucial negative role.

But first, this chapter and the next introduce the phenomenon and concepts involved in describing, estimating and managing the adverse effects of *heterogeneity* in sampling. Detailed treatment of all sampling errors will gradually be developed in later chapters, but an appropriate initiation can also be embedded in the story of heterogeneity starting right here.

Heterogeneity: essential features
The reader is referred back to Figure 1.2 for a generic illustration of key elements of the concept of heterogeneity, here in a distinctly segregated material. Observe how heterogeneity dictates the possibilities of sampling from the entire lot: grab sampling vs composite sampling, which will be the key theme throughout this book. Clearly a drill core (considered as a stack of individual increments) securing a complete top-to-bottom column will be a particularly desirable composite sample.

Figure 1.2. The arch-enemy of representative sampling: *heterogeneity*. A "grab sample" (a single event sampling) can never be representative of a heterogeneous lot.

CSE: Correct Sampling Errors
FSE: Fundamental Sampling Error
GMO: Genetically Modified Organism
GSE: Grouping and Segregation Error
DH: Distributional Heterogeneity
ISE: Incorrect Sampling Errors
IDE: Increment Delimitation Error
IEE: Increment Extraction Error
IWE: Increment Weighing Error
IPE: Increment Preparation Error
TSE: Total Sampling Error
TAE: Total Analytical Error

3.1 Introduction to the concept of sampling errors (excerpt from DS3077)

3.1.1 Correct Sampling Errors (CSE)

3.1.1.1 Fundamental Sampling Error (FSE)

The Fundamental Sampling Error (FSE) is always present in all sampling operations. Indeed, even a fully representative, non-biased sampling process will be unable to materialise two samples with exactly the same composition, *because* the lot material is heterogeneous. FSE is the smallest possible sampling error obtainable for a material system in a given grain size distribution state. FSE can still be of significant magnitude even in the absence of all other sampling errors, especially if the analyte is present only in trace amounts spatially distributed in an *irregular* fashion (the very definition of heterogeneity). This is the case, for example, for trace constituents, contaminants and genetically modified organisms (GMOs). FSE is the minimum unavoidable reflection of the difficulties encountered when a minute volume/mass sample is used to characterise the properties of a complete lot made up of heterogeneous material. FSE is reflecting what is termed the compositional heterogeneity, CH (see Chapter 4 for details).

3.1.1.2 Grouping and Segregation Error (GSE)

GSE plays a significant role in addition to FSE. GSE originates from the inherent tendency of lot units, grains, particles to segregate and/or to group together (local spatial coherence), to varying degrees within the lot volume. This kind of irregularity typically manifests itself differently at different scales, strongly influenced by grain-size heterogeneities as well—the full complement of this spatial irregularity is termed Distributional Heterogeneity, DH (see Chapter 4 for details). Unlike FSE, the effects from GSE can be reduced by certain operations, e.g. by mixing, which can reduce segregation and grouping effects significantly, although never eliminate them completely (Figures 3.1 and 3.2).

Heterogeneity—The Root of All Evil 1

Figure 3.1. Example of typical manifestations of heterogeneous material in the laboratory, here herring fillets subjected to sample processing and preparation in a mixer. The laboratory technician may believe that the resulting "homogenate" (right) is sufficiently well comminuted and mixed to allow direct aliquoting with a spatula (grab sampling), extracting only the precise, very small amount needed for analysis. The "homogeneity" is routinely assessed by visual inspection only. However, this is a major fallacy, as shown in Figure 3.2.

Figure 3.2. Significant compositional heterogeneity at the end state of "thorough mixing" of a batch of herring fillets in the mixer shown in Figure 3.1. Hitherto *invisible* compositional differences can be seen by high-powered illumination combined with a UV camera-filter, revealing an appreciable "hidden" residual heterogeneity in what is normally called the "homogenate". Compositional Heterogeneity (CH) and Distributional Heterogeneity (DH) are further discussed in this and the following chapters.

All illustrations in this book have been selected also to act as generic examples from a much broader range of applications. It is only the material heterogeneity that matters when it comes to sampling.

3.1.2 Incorrect Sampling Errors (ISE)

3.1.2.1 *Increment Delimitation Error (IDE)*

IDE reflects an increment delimitation problem associated with sampling in practice. IDE occurs when the boundaries of an *intended* increment cannot be assured to be identical to those for other increments. For example, for 1-D sampling, an increment shall be delineated by parallel boundaries and shall provide a complete cross-section of the moving flux of matter, i.e. covering both transverse dimensions (width, thickness) orthogonal or oblique to the transportation direction. IDE occurs when all parts of an intended, delineated increment do not have an exactly identical chance of becoming part of what is actually extracted.

3.1.2.2 *Increment Extraction Error (IEE)*

IEE occurs when the sampling tool is selective on what is extracted. For example, particles hitting the boundary wall of the increment tool must obey the so-called centre-of-gravity rule: particles having their centre-of-gravity inside the delimited tool boundaries must be intercepted so as to be included in the increment—and vice versa for particles for which the centre-of-gravity falls on the outside (such particles must not be included). IEE reflects a practical increment recovery, or extraction, problem, and the solutions needed to suppress IEE are intimately associated with the design, manufacturing, operation and maintenance of the specific sampling tool. IDE/IEE can be viewed as two aspects of the same practical issue. IDE/IEE are very often the main culprits in unwanted sampling errors.

3.1.2.3 *Increment Weighing Error (IWE)*

IWE reflects specific weighing uncertainties or, for process sampling, when all collected increments are not proportional to the contemporary flow rate (1-dimensional lot) or to the thickness of an elongated stratum (2-dimensional lot) at the time or place of collection. IWE can often be dealt with rather easily, but not always.

3.1.2.4   Increment Preparation Error (IPE)

IPE reflects any-and-all post-sampling alterations that may occur to an already extracted increment or sample, e.g. as a result of contamination, spillage, losses of moisture, alteration (physical constitution or chemical composition), human errors (ignorance, carelessness, fraud or sabotage). IPE is strictly speaking not a sampling issue, as its effects only occur *after* sampling operations. However, there is good reason to categorise IPE together with the other ISE, as the resulting effects add to the total sampling error (TSE) in a similar way before analysis.

3.2   Heterogeneity—the basics

Heterogeneity is responsible for the effects of the two "Correct Sampling Errors" (CSE), and its interaction with the sampling process leads to effects from four additional "Incorrect Sampling Errors" (ISE). It is necessary for all competent samplers to have a basic grasp of the nature and manifestations of heterogeneity, in order to be able to assess the appropriateness of existing sampling procedures and equipment. There are infinitely many manifestations of heterogeneity, yet for the competent sampler there is only one issue at hand: representative sampling is nothing but *heterogeneity-counteracting* mass reduction, Figure 3.2.

3.3   Materials, sampling targets and lots

Materials (sampling targets and lots) present themselves in a bewildering array of different types and degrees of heterogeneity with many diverse physical manifestations, Figures 3.3 and 3.4. Materials may appear as discontinuous or continuous solid(s), as

Figure 3.3. The many varied manifestations of compositional and distributional heterogeneity—no homogeneous materials exist in the world of science, technology and industry. Consequently, the path from lot to aliquot is the most critical part, because this gives rise to the total sampling error (TSE)—very often dominating over TAE. There are very many examples of heterogeneous materials. This figure is only an attempt to present *some* typical materials, which will be complemented throughout the book. For the novice reader it may well be far from obvious that a singular approach exists for how to sample such manifestly different materials. But read on with confidence, the TOS is here!

discrete materials composed of varying types of mixtures of component units (particles, fragments, other types of components), aggregates, two-phase systems (e.g. slurries...) or three-phase systems (solids, liquids, gases). Examples of compositionally heterogeneous materials are legion, the examples shown here are but a very few illustrations of the broad range of potential sampling targets of interest. Distributional Heterogeneity (DH) characterises the scale regime between increments and the full lot; process

Figure 3.4. Heterogeneity will always interact with sampling and will determine the degree of TOS-compliant work needed to *counteract* its negative effort, which gives rise to sampling errors. Top left: grab sampling (discrete sampling) of highly heterogeneous slurry (grapes/must) at a winery intake. Top right: array of optional increment sizes for sampling of soil with intermediate heterogeneity. Lower left: sampling targets, e.g. as "big bags", offer the added difficulty of the sampler not being able to observe the material and its heterogeneity (but it certainly exists). Lower right: manual process sampling (grab sampling) of apparently uniform material. The heterogeneity manifestations shown cover the range from high to low, from visible to hidden, from the considered to the neglected. N.B. samplers shown here are **only** illustrating and acting as grab samplers in the name of TOS education!

sampling (dynamic 1-D lots) will be covered later. The present descriptions and examples focus on the *generic* aspect of heterogeneity and its interaction with the sampling process; the examples should be easily translated into all typical types of material(s) of particular interest to the reader.

It is one of the most powerful features of the TOS that it offers *universal principles* for representative sampling that cover all materials and their manifestations of heterogeneity. A first lesson is that, while dramatically different in their *apparent* physical manifestations, all materials present *identical* heterogeneity challenges, which **only** differ in degree; therefore, they are treated in identical fashion by the TOS. This is a tremendously liberating insight. Also, once we know how to deal with heterogeneity on one particular scale, we can deal with all manner of lots—for all materials, under all conditions, at all scales.

Many meso- and large-scale heterogeneity manifestations are *deterministic*, in that they result from specific processes, e.g. manufacturing/processing, stock laying-up processes, transport and pouring processes, flow processes. This very nearly always involves some form of dynamic activity, i.e. heterogeneous three-dimensional sampling targets are nearly always temporarily present in a moving one-dimensional configuration (flowing, ducted, conveyed, transported). Such sampling targets can very easily be sampled, e.g. by way of *interception* by a cross-cutting sampling tool; much more of this "easing" below.

3.4 Homogeneity–heterogeneity

For the purpose of a precise understanding of the concept of heterogeneity, it is necessary to present a strict definition of its opposite: *homogeneity*, or rather of what constitutes a *homogenous material*. Several definitions abound in the general literature, but we here focus on the most stringent one (offered by the

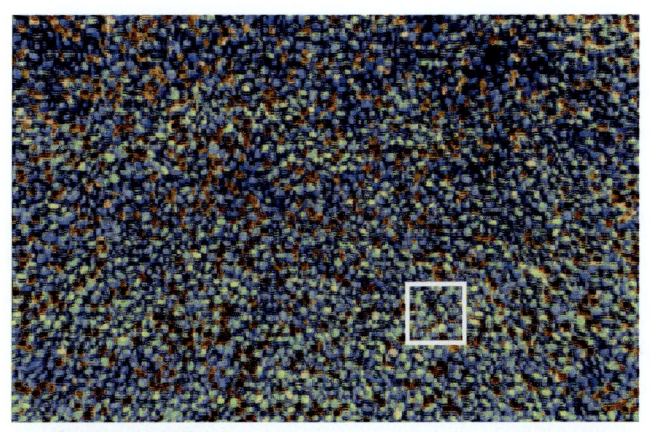

A - Grab sampling

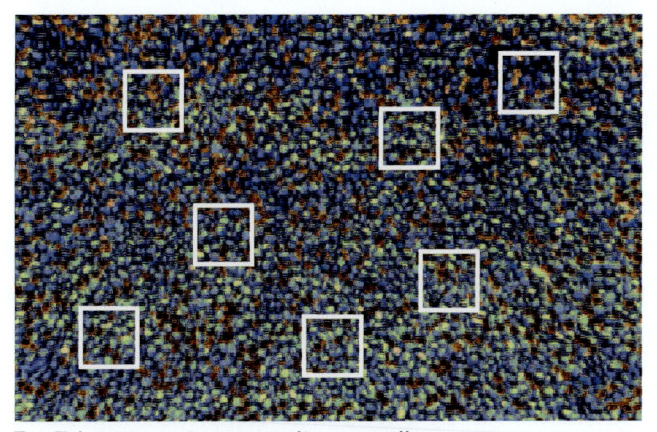

B - 7-increment composite sampling

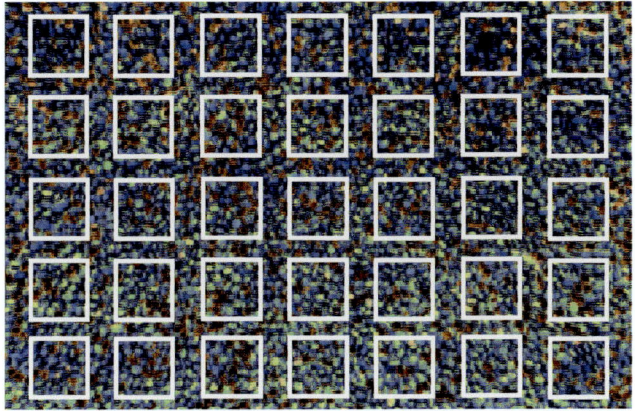

C - 35-increment composite sampling

Figure 3.5. Significantly heterogeneous material, as sampled wrongly (A) and with a much better chance of being representative (B). Note that the "overkill" 35-increment example (C) is only used to emphasise the point that composite sampling needs to make use of the optimal, i.e. the necessary and *sufficient* number of increments, Q, with which to counteract the specific heterogeneity met with. Finding Q is one of the two major objectives of representative sampling.

> **Constituent units**
>
> Regardless of which analyte is of interest, any constituent unit will be characterised by a certain quantity thereof, and in this quantitative measure, the concentration, may vary between 0% and 100%. The TOS' tradition contains an empowering twist in which grains, as well as their possible fragments (fragmented due to the sampling process), are **all** termed "fragments". From a generalisation point of view, it will be seen to be very convenient to term both the original unaffected grains as well all possible fragment cascades induced in/by the sampling process itself *generically* as fragments. This makes it possible to deal with all types of original materials and their undisturbed constituent units, at all scales up to the full scale of the target, as well as those sub-parts, which are now made up of fragmented grains. In this manner, one can speak with complete generality of, and deal conceptually with, *any* type of sampling target, which is then, to the first conceptual consideration, simply made up of *fragments* in more or less complicated spatial aggregations.

TOS): a homogeneous material is composed of *strictly identical units (grains, fragments ...)*—identical in the most comprehensive sense, i.e. *all units* must be of the *exact* same size, composition, density, surface morphology, features (e.g. wettability) and electrical charge (such differences will lead to differential segregation or flow effects between fragments). It is clear that having *strictly* identical units is a very strong requirement that leaves almost no candidate in the real world of naturally occurring, manufactured or processed materials that is homogeneous. In this context, it is best and indeed safest, always simply to *assume* that *all* materials that are to be sampled are heterogeneous. This is a sound scientific attitude that will ensure that heterogeneity is always the most important item on the sampling agenda. It really does not matter whether there *might* be a few exceptions from this general rule. Instead of wasting valuable time and effort on extensive testing *if* this were indeed the case for an extremely rare candidate material, by treating all materials identically as if they were indeed all significantly heterogeneous, the way is open for truly empowering simplicity: the TOS offers a *unifying* set of principles and unit operations with which to treat all cases.

References 1 and 2 offer a more comprehensive treatment, ideal to build on the present introduction. It is of no consequence for the reader that it nominally addresses the food, feed and GMO sectors, in fact it is universal for all types of material, and is strongly recommended.

3.5 Scale

It is necessary first to focus on the *inherent* heterogeneity of lots (Compositional Heterogeneity, CH), which turns out to be understandable from only three concepts:
i) constituent "units" (of various kinds),
ii) three operative scale levels and

iii) simple summary statistics (average, standard deviation, variance). Crucially, this application of statistics does **not** address analytical results—more on this issue below.

The scale levels mentioned above are also known as "observation scales" in the TOS literature, sometimes also referred to as "observation volumes". Thus, all materials are made up of "constituent units" on three scales, e.g. starting from the absolutely smallest scale:

- atoms and molecules, but this scale level is generally not of interest for sampling in technology, industry and society;
- the critical scale level commensurate with the sampling tool volume in which the constituent units would be: grains, particles, fragments and coherent aggregations (coherent enough so as not to be fragmented in the sampling process);
- the largest scale of interest is the observation scale corresponding to the sampling target itself (the lot scale).

This three-tiered scale hierarchy constitutes the essential scaffolding for the TOS' theoretical and practical concepts regarding heterogeneity; nothing more is needed.

3.6 Heterogeneity vs sampling

All materials to be sampled are compositionally heterogeneous, because all fragments most certainly cannot be assumed to carry an identical concentration of the analyte in question, and thus neither an equal "amount of heterogeneity". Therefore, a significant compositional heterogeneity will be present. It is of no consequence if only a few, or an overwhelming proportion, of the fragments turn out to be identical, or different, in practice; the material is still heterogeneous—or think of a material consisting primarily of a uniform set of grains but *contaminated* with trace amounts of an extraneous (or intrinsic) analyte, which could be present in/as grains of different size, or which could be present as precipitations on grain surfaces. Such sampling targets present more difficult cases to deal with, because the heterogeneity reflects a necessarily irregular *spatial* distribution of the sparse units carrying the contaminant. Thus, there is also a significant DH.

Most materials also display a non-uniform distribution of grain sizes (very few materials are truly mono-disperse), in which case the constituent units differ both with respect to their size, volume and mass as well as their analyte concentrations. This type of heterogeneity can be said to be a *structural property* of the material. Theoretically the TOS considers grain-size heterogeneity as part of the CH.

The topic of heterogeneity is treated in more depth in Reference 2, wherein can also be found a plethora of further background references.

3.7 References

1. *J. AOAC Int.* **98(2)** (2015). https://www.ingentaconnect.com/content/aoac/jaoac/2015/00000098/00000002, bit.ly/tos2-5
2. K.H. Esbensen, "Materials properties: heterogeneity and appropriate sampling modes", *J. AOAC Int.* **98(2)**, 269–274 (2015). https://doi.org/10.5740/jaoacint.14-234, bit.ly/tos1-5

4 Heterogeneity—the root of all evil (part 2)

All sampling has to deal with materials that are heterogeneous at one scale or another (or at all scales). Whatever sampling procedure to be used, its primary objective is to *counteract* the effects of the material heterogeneity. Up to six sampling errors are potentially in play, because sampling processes necessarily *interact* with heterogeneous materials. Two sampling errors arise because of lot heterogeneity (Correct Sampling Errors, CSE), and four additional errors are produced by the sampling process itself (Incorrect Sampling Errors, ISE) if not properly recognised and reduced substantially or eliminated completely. This chapter leads the reader further into the concepts used in describing, estimating and managing the adverse effects of heterogeneity in sampling.

AA
CSE: Correct Sampling Errors
ISE: Incorrect Sampling Errors
CH: Compositional Heterogeneity
DH: Distributional Heterogeneity
a_L: average lot concentration
GMO: Genetically Modified Organism
FSE: Fundamental Sampling Error
GSE: Grouping and Segregation Error
FSP: Fundamental Sampling Principle

4.1 Introduction

Compositional heterogeneity (CH) is a reflection of the intrinsic compositional differences *between* the ensemble of individual units which make up all lots (grains, particles, fragments, other). A material will display a non-zero constitutional heterogeneity whenever it is made up of *different* constituent units. Mixing will have no effect on this type of heterogeneity. It will be the exact same ensemble of units regardless of to what degree they are mixed up—they remain equally different.

The TOS has coined the concept *heterogeneity contribution* for the contribution made to the full lot heterogeneity by each individual fragment (CH scale) or an increment (DH scale, see further below). It is advantageous to focus on heterogeneity contributions because this allows the individual fragment masses to be factored in. Large fragments (defined as masses larger than the average fragment mass) may also carry a large concentration deviation from the average lot concentration, a_L, with the consequence that the heterogeneity contribution from such a fragment will be large. However, should a fragment, identically large in size (mass), happen to have a concentration very close to, or perhaps even (accidentally) *equal* to, a_L, its contribution to the full lot heterogeneity will be insignificant, regardless of its mass; it is simply a large fragment with (almost) precisely the average lot composition.

Had such a fragment been *grabbed* from the lot by accident, its analytical result would have been both accurate and precise, indeed representative of the whole lot—alas, such miraculous knowledge is **never** known at the time of sampling.

From the perspective of fragments with significant compositional deviation, large fragments will always contribute overwhelmingly to CH_L, while individual small particles (grains of dust for example), will not matter much to the total material heterogeneity because of their infinitesimal weight. Collectively, however, the fine fraction of a lot (depending upon this collective mass proportion) *may* contribute appreciably to CH_L if the composition of the units in this fraction deviates significantly from a_L.

It follows that the visual *appearance* of a lot made up of an array of discernible fragments may very well give a false, or only a superficial, impression of the state of heterogeneity, because large fragments will dominate the visual grain size distribution expression. However, the human eye will not, in general, be able to *see* the analyte concentrations involved. Similarly, a material made up of almost identical grains, e.g. cement, ground coffee, soy beans, wheat grains, sugar, "fines" (the latter because of their similar scattering effects) etc. may nevertheless have a highly significant heterogeneity contribution, e.g. regarding trace concentration analytes (e.g. toxicants, mycotoxins or GMOs). Observe that a material may simultaneously appear very close to homogenous (e.g. sugar or, say, a 99.9% pure chemical compound), while in reality representing a very large heterogeneity with respect to an impurity analyte, which necessarily must be distributed very irregularly at such low concentrations (Figure 4.1). There is a powerful lesson to be learned from these simple relationships: the visual impression of heterogeneity can be grossly misleading—and because one will never know if this is the case, or not, the visual impression must consequently **never** be used as a basis for "homogeneity assessment". Indeed, the practice of speaking of "homogeneity" can just as well be abandoned—when it comes to sampling, it is **all** about the degree of heterogeneity.

Figure 4.1. "Homogeneous powder" with a normal grain-size distribution, but the larger-than-average particle sizes have been dyed blue, allowing detailed insight into grain-size differentiation and segregation behaviour. The original powder visually makes a totally homogeneous impression ("white powder"). Two different DH_L manifestations are shown to the left and centre. The latter was produced by a single 90° rotation around the vertical axis of the container, illustrating that DH_L manifestations are often transient phenomena, which are a sensitive function of a number of agitation factors active in production, handling, transportation and by manipulation while being sampled. The constant presence of the gravitational force field and/or the transient presence of centrifugal forces (as in the example above) will impact the specific DH manifestation of very many types of aggregate materials. The right-hand photo shows the important effect of *pouring segregation* and the resulting problems in trying to acquire a representative single-sample aliquot using a laboratory spatula. Such discrete sampling operations (grab sampling) can never be representative. Original photos courtesy of Peter Paasch-Mortensen (reproduced with permission).

Digging into the mathematical manifestation of the TOS just a little, here follow the proper statistical definitions of CH and DH:

4.2 Constitutional Heterogeneity (CH)

The TOS defines a *heterogeneity contribution* to the total lot heterogeneity by first focusing on the scale of the individual fragments (grains). The TOS characterises all fragments according to the analyte of interest, A, expressed as a proportion (or grade), a_i, and the fragment mass, M_i. If a lot consists of N_F individual fragments with individual masses, M_i, which together display an average fragment mass $M_{\overline{i}}$ and with lot grade designated a_L and the lot mass M_L, the *heterogeneity contribution* from each individual fragment, h_i, is given by Equation 4.1:

$$h_i = \frac{(a_i - a_L)}{a_L} \cdot \frac{M_i}{M_{\overline{i}}} = N_F \frac{(a_i - a_L)}{a_L} \cdot \frac{M_i}{M_L} \qquad (4.1)$$

It will be seen that heterogeneity contributions are dimension-less intensive quantifiers. h_i expresses both the compositional deviations of each fragment, while also factoring in variation in fragment masses.

This viewpoint constitutes a major distinction from "classical statistics" where all population units contribute equally ("with equal statistical mass"). h_i constitutes an appropriate measure of mass-weighed heterogeneity as contributed by each of the N_F fragments, which together make up an entire heterogeneous lot.

Based on this definition, the **total** constitutional heterogeneity of the lot, CH_L, can easily be defined on the basis of the individual h_i's—as the variance of the distribution of the heterogeneity contributions of **all** fragments in the lot (Equation 4.2):

$$CH_L = s^2(h_i) = \frac{1}{N_F} \sum_i h_i^2 = N_F \sum_i \frac{(a_i - a_L)^2}{a_L^2} \cdot \frac{M_i^2}{M_L^2} \qquad (4.2)$$

This variance measure is a convenient estimate of the total heterogeneity variability in the lot. The spatial analyte distribution of a heterogeneous lot does not have to comply with a random distribution assumption, in fact it rarely does, as discussed below. The TOS has nevertheless derived the above theoretical understanding of CH_L, resulting in the practical and very simple first view that all lots are made up of the sum of N_F specific heterogeneity contributions from each fragment.

4.3 Distributional Heterogeneity (DH)

By ascending one scale level, from the scale of fragments to the operative level of one sampling unit (sampling scoop), the *increment*, it is possible to address the complementary realm of the spatial lot distributional heterogeneity, DH_L. No longer concerned with the lot consisting of the totality of N_F fragments, at this larger scale of scrutiny, lots can alternatively be considered as being made up of a number of *potential* sampling increments (in the TOS termed "groups" or "groups-of-fragments"), N_G, commensurate with the selected operative volume of the sampling tool in use.

Other than this hierarchical scale difference, the formalism is identical; i.e. we are still interested in a quantitative description of the differences in composition (concentration of the analyte, a_n) but now between the groups (increments). DH_L is calculated in a strict analogue to the definition of heterogeneity carried by a single fragment, only now the effective unit is the *sampling increment*, which will contain a group-of-fragments that will be characterised *in toto*. Thus, a *group* in the lot (index n), G_n, similarly carries a contribution of the total lot heterogeneity, h_n, which can be calculated from the analyte concentration in the group in question, a_n, the group mass, M_n, the average group mass $M_{\bar{n}}$ and the average grade over all groups, $a_{\bar{n}}$ as defined in Equation 4.3.

$$h_n = \frac{(a_n - a_L)}{a_L} \cdot \frac{M_n}{M_{\overline{n}}} = N_G \frac{(a_n - a_L)}{a_L} \cdot \frac{M_n}{M_L} \qquad (4.3)$$

The **total** heterogeneity for the entire lot can similarly be calculated in an identical fashion as for fragments, as the variance of **all** group heterogeneity contributions (Equation 4.4):

$$DH_L = s^2(h_n) = \frac{1}{N_G} \sum_n h_n^2 = N_G \sum_n \frac{(a_n - a_L)^2}{a_L^2} \cdot \frac{M_n^2}{M_L^2} \qquad (4.4)$$

It is observed that both CH_L and DH_L are defined by mass-weighted analytical composition deviations.

But there is more to the latter DH_L than appear from a mathematically identical formalism. This is where the founder of the TOS made a creative, indeed ingenious interpretational leap.

Due to the fact that the aggregate sum of all (virtual) groups constitutes the complete lot in its geometric entirety, it follows that DH_L is, in fact, also a measure of the total *spatial heterogeneity* exhibited by the lot, hence the term "distributional heterogeneity", DH_L. From this complementary point-of-view (at the scale of groups), the lot is made up of a set of virtual groups, N_G in total. This dual-scale understanding of the heterogeneity of any lot (system, material, process stream) constitutes a surprisingly powerful theoretical concept with which one is able to understand all key issues of heterogeneity and how this sets up the conditions for representative sampling. DH_L accounts for the material lot heterogeneity in a particularly relevant form, namely that corresponding to the specific sampling tool size used, characterised by a specific increment mass, M_S. The effective group in question is simply the group-of-fragments that resides *in the sampling increment used*.

Thus, it is possible to ascertain the quantitative effect of the lot heterogeneity *interacting* with alternative sampling processes, for example using alternative sampling tool volumes or by using an alternative sampling procedure. This would result in numerically *different* measures of the spatial heterogeneity.

> This relationship cannot be stressed enough. Estimation of the intrinsic heterogeneity of any lot can **only** be carried out based on one (or other) sampling tool volume, or increment size, as employed in one, or other, specific sampling procedure. The issue of heterogeneity estimation is, both in principle as well as in practice, a *vicious circle*. A specific sampling procedure must be used in order to be able to estimate the sampling variation due to the heterogeneity—but all sampling procedures are always influenced significantly by the very same heterogeneity which is to be characterised.

There is an understandable wish, from people not experienced with the TOS, that one *"should be able"* to calculate an optimal increment size, for example for engineering purposes or similar. What is not completely thought through in such *"demands"* is that both the size of an increment, as well as the total number of such increments needed for composite sampling in order to counteract a particular lot heterogeneity, must logically and very much in practice, be based on a valid estimate of said lot heterogeneity in the first place—which, however, necessitates that a specific increment size has already been chosen, see box above. This type of demand thus sets up a vicious circular argument that has to be broken. This is often done by a very practical approach in which a *tentative* increment size (and a corresponding total number of increments to form composite samples, Q) is first tried out, which will lead to a certain total sampling variability as expressed by replicated sampling (for details see Chapter 9). If this variability is too large with respect to some *a priori* defined threshold… something has to give, and this is either Q (which will have to be increased) and/or the increment size needs to be adjusted. More of these key issues will be discussed below. This perhaps haphazard-looking procedure is in reality very often guided by quite specific knowledge and experience as to the material in question, making the choice of increment size (and Q) much less of a guess; for example by making these two parameters at first corresponding to what is the current situation. Based on this, the resulting empirical sampling variability can then either be acceptable or not—and only in the latter case will it be necessary to modify the sampling procedure.

A specific type of increment, termed the proverbial *grab sample* has historically been very much used, particularly in industrial applications. The most important aspect of any such single-increment sampling process is the size of this sampling unit M_S and the way it has come about. But as will become compellingly clear, such a single-scoop sample is very, very rarely acceptable (it results in an unacceptably large sampling bias), so M_S is in fact very nearly always to be understood as the *compound mass* of a *composite sample*, made up by Q increments. One of the most important objectives of the TOS is to allow the sampler the most reliable way to estimate the magnitude of Q, see, for example, DS 3077.[1]

4.3.1 Action helps

Unlike CH_L, which is an intrinsic characteristic of the given material, DH_L can actively be altered (reduced), either by choosing a smaller sampling tool (thereby increasing the number of increments in composite sampling—in process sampling this means increasing the sampling frequency) and/or the lot can be thoroughly *mixed*, blended etc. In large lots, forced mixing is often impractical or impossible, however; in such cases increasing the number of increments is the only option for more reliable primary sampling. If there is a significant segregation or grouping (or fragment clustering) in the lot, increasing the sample size for a one-increment grab sample will only result in a very minor effect and will soon reach an impractical limit making the sample way too big. By way of contrast and effectiveness, composite sampling is *always* a good choice of action. It is advantageous to think of more increments as synonymous with a sure way to secure a better, a more fully deployed geometrical lot *coverage*.

It follows that sampling from a heterogeneous lot can never result in completely identical analytical results; there will always be a sampling variance (more accurately, a sampling-and-analysis variance) as expressed by the empirical differences between a set of analytical results. Even a set of identically replicated samples, carried out following an identical laboratory protocol, will give rise to a distinct, non-zero sampling variance. This is solely due to the fact that no sampling process can fully eliminate the effect of heterogeneity at all possible scales (FSE, GSE).

The role of representative sampling is to reduce these fundamental sampling effects as much as possible, and to be able to quantify the remaining sampling variance. It *may* happen that particular systems possess extraordinarily small heterogeneities, but no generalisations as to universal relationships regarding "homogeneity" or "sufficient homogeneity" can be drawn from such particular instances. It is imperative always to treat **any** lot material as if it carried a significant degree of heterogeneity

4.4   Heterogeneity vs practical sampling

If one was obliged to produce a *complete* heterogeneity characterisation of a specific sampling material, it would be necessary to analyse, and weigh, **all** constituent fragments. This is obviously not possible, nor desirable in sampling practice. Because of this impossibility, *sampling* comes to the fore: only a part of the lot will be physically sampled and eventually analysed.

4.4.1 Enter the sampling increment

What constitutes an *ideal sample*? An ideal sample would have to be composed of a subset of single fragments selected *individually* from the lot, completely *at random*, i.e. based on total free access to the full geometrical target volume. This latter demand is codified in the TOS' Fundamental Sampling Principle (FSP): there must be free access to absolutely every grain (fragment) in the sampling process, no exception.

But it is clear that nobody in their right mind would wish to collect an ideal sample in this sense *in practice*. All samplers must instead accept that one can only sample a set of coherent neighbouring fragments, a group-of-fragments, a set of *increments*. While any size and disposition of a group can be envisaged, the TOS is in practice **only** interested in the special group-of-fragments that will end up **in** the sampling tool after an incremental sampling operation. Thus, for very practical reasons, the TOS is only interested in those (theoretical) *groups* that make up extracted *increments* (in practice). Sampling, therefore, in practice always takes place by extraction of increments of a size that needs to be optimised.[1]

4.4.2 There are lots, *and* there are lots...

This understanding points to two alternative views of any sampling target, see Figure 4.2. This division establishes the only two scales-of-interest when coming to grips with the complexities of heterogeneity, the fragment scale and the increment scale. There is actually only one more scale of interest, the full lot size (mass), M_{Lot}. There is great conceptual and practical power in these simple scale relationships.

Before sampling, only *virtual groups* exist in the target lot. Even though viewed from the point of view of the increment scale, the entire lot ensemble of different fragments is still available for manipulation, so mixing will have an effect on the compositional

N_F fragments **N_G groups (groups-of-fragments)**

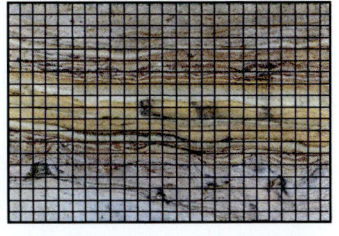

Composite sample (10 increments) **Composite sample (50 increments)**

Please refer to Figure 5 in *J. AOAC Int.* **98(2)**, 269–274 (2015). https://doi.org/10.5740/jaoacint.14-234 for a complementary presentation of Pierre Gy's brilliant distinction between fragment and group-of-fragments (increments).

Figure 4.2. Pierre Gy's inspired conceptual scale-jump from fragment-scale to group-scale, which allows definition of DH (see text). The different observation scales correspond to the different sampling units indicated, fragment vs group. All sampling targets (lots) can be viewed from these alternative vantage points, i.e. either as a collection of N_F fragments or a collection of N_G groups (increments). The third scale level in the TOS corresponds to the sampling target (lot) itself. The illustration portrays a geological section through a layered sedimentary sequence displaying considerable lateral and horizontal heterogeneity. Sampling this lot with a "too-small" number of increments would have severe adverse consequences. Illustration courtesy of Mr Martin Lischka, reproduced with permission.

differences *between* virtual groups. Consider an operation like forceful mixing: shaking a pitcher with different types of solid particles; mixing of a segregated slurry tank; whisking together egg white and yolk; shaking a cocktail; or mixing in blenders. It is easy to picture in the mind's eye how fragments become increasingly spatially mixed from such an operation. Mixing has a definitive influence on between-group differences, as virtual groups come to be more and more *similar* as mixing progresses. Note, however, that while the DH of the lot is reduced, lot CH remains the same.

Note also that there is a limit to this operation: "infinite" mixing will **not** lead to a homogeneous material state, but only to a *minimum residual heterogeneity* state, after which more mixing only results in a mixing/de-mixing *steady-state* with random deviations confined to oscillate around this state (it may even increase DH locally). Close to this state, there is thus a definite limit to how much good mixing can do—more mixing will far from always lead to a lesser heterogeneous result. What follows is a glimpse into the more theoretical deliberations within the TOS with a view of indicating its theoretical rigor and power.[a]

4.4.3 Key interrelationships in more depth

Within the TOS there exists an intimate interrelationship between the Grouping and Segregation Error (GSE) and the Fundamental Sampling Error (FSE), expressed via the segregation factor, ζ and the grouping factor, γ, Figure 4.3. These phenomenological factors are used in the TOS' theoretical treatment of heterogeneity to *represent* segregation and local "grouping" in the more comprehensive equations originally derived by Gy. While it lies outside of the scope of this introduction to present the theoretical treatment in full, one of the factors simply turns out to represent the operative size of the increment used for sampling, for which reason it is easily appreciated how important a role γ plays for practical sampling. And the other, the segregation factor ζ, lends itself well to direct graphical illustration and understanding. Figure 4.3 thus shows with clarity the effect of reducing the Segregation factor as a function of gradually increased mixing.

---

[a] It is **not** necessary to be able to comprehend these theoretical relationships in full at this introductory stage; it is enough simply to "follow the finger that shows" (Heidegger).

$$s^2(GSE) = \zeta \times \gamma \times s^2(FSE)$$

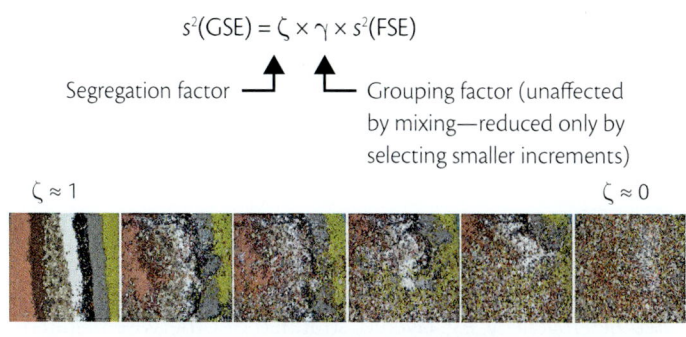

Segregation factor ↑ ↑ Grouping factor (unaffected by mixing—reduced only by selecting smaller increments)

$\zeta \approx 1$ $\zeta \approx 0$

Figure 4.3. Illustration of the counteracting effect of mixing significantly heterogeneous material, together with a theoretical equation of the relationships between four elements in the TOS. Progressive mixing is tantamount to reducing the segregation factor (ζ) closer and closer to 0 but never reaching it completely (which would represent a homogeneous mix, a physical impossibility). For more detail see, for example, References 1–5.

Upon reflection, it will be appreciated how the GSE effects can be reduced by reducing either one, two or all three of the factors on the right-hand side of the equation. Observe how only the segregation factor represents mixing, while the grouping factor represents a pre-requisite for effective composite sampling—reducing the size of the operative increments.

Without going into further details here, this key equation in fact makes it possible to derive no less than three of the four sampling unit operations: composite sampling, crushing (comminution) and mixing.[1]

While CH_L resides in the scale interregnum between fragment and the lot, DH_L quantifies the heterogeneity that can be attributed to the realm between the increment scale and the full lot size. Both these heterogeneity aspects are needed to fully characterise the total heterogeneity of a material, but they cannot be physically separated from one-another. CH_L and DH_L are conceptual, theoretical components with their individual physical meanings that, in practice, exist simultaneously for any physical material. Not surprisingly, from their closely related physical meanings, there is also a close mathematical relationship between CH_L and DH_L as is illustrated in Figure 4.3.

4.5 "Structured heterogeneity"

Many materials display heterogeneities with a special twist, *structured heterogeneity*, e.g. layered, stratified or otherwise hierarchically organised heterogeneities (Figure 4.4).

Figure 4.5 is an example from the food and feed sectors, in which the food elements (minced meat, spices and fat) are present in a very irregular relationship, which is indeed the reason behind this particular type of the well-known brand of sausage called

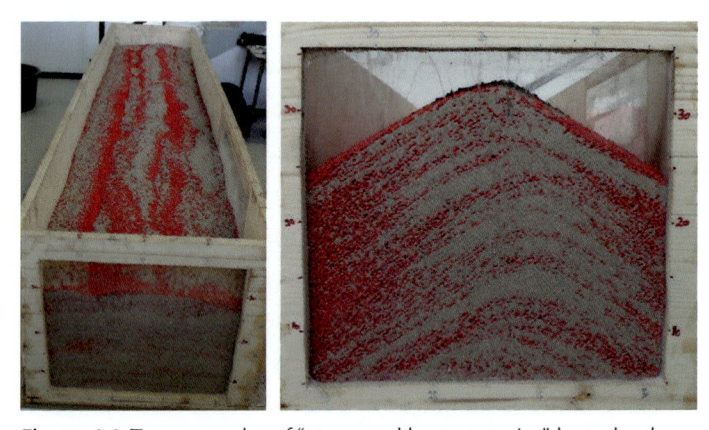

Figure 4.4. Two examples of "structured heterogeneity", brought about by different stacking processes, illustrating the type of spatial heterogeneity, DH_L, often present in transportation or storage depots, vessels, trucks, train loads, containers, ship cargo holds and sometimes in reaction and production vessels as well. Both lot examples have identical CH_L, portraying a 10% "analyte" (red plastic pellets) in a 90% sand matrix. Partially structured heterogeneity is a characteristic that occurs often in technology and industry, due to the extensive stacking, reclaiming and transportation processes involved in production, processing and manufacturing. This shows very clearly why grab sampling always comes up short also against materials with structured heterogeneity—as indeed against all materials with a significant DH_L.

Figure 4.5. Salami ("Chorizo") can be viewed as an example of highly irregular but still partially structured material heterogeneity. Tube-coring (using a sampling thief) will not necessarily guarantee a representative sample. Observe, however, how the principle of riffle-splitting easily can be applied to this material allowing it to be handled in a fashion identical to that of free-flowing aggregate material, see Chapter 14. Riffle-splitting slicing can often be conducted so as to guarantee representative sub-sampling, see Petersen et al.[2]

"Chorizo". Sampling such a material, for example for food compliance or safety purposes, is not a straight-forward issue, perhaps contrary to common thought: "How difficult can it be to sample a salami-type sausage in the analytical laboratory?" In many sectors, sampling of material with this and *similar* partially structured heterogeneities takes place with a tubular corer (sampling thief, sampling spear of appropriate dimensions), but from the above comments it should be clear that a core section of this material runs a risk of being non-representative, see also Chapter 2.

An alternative approach is inspired from the principles of *riffle-splitting*, see Figure 4.5. By slicing a sufficiently high number of slices, each covering a full cross-section of the lot, a division of the material in all aspects identical to riffle-splitting is obtainable. By selecting a set of slices of equal thickness, a correct TOS mass-reduction can be achieved even for materials with such

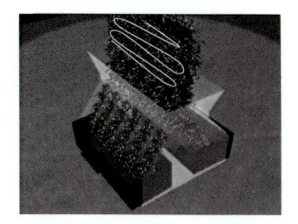

Figure 14.3. Longitudinal loading of the ingoing sample to be split is often an area of major misunderstanding.

very high CH_L and DH_L.[2] Despite the apparent heterogeneity difficulties, a 100 % TOS-compliant mass reduction can be achieved with ease (albeit often necessitating *some* practical work). The effectiveness, i.e. the representativeness of sub-sampling is actually only a matter of practical implementation, i.e. how many thin-slicing cuts one is willing to produce and how many of these one is willing to aggregate as composite samples. More about riffle splitting and laboratory mass reduction (sub-sampling) in Chapter 14.

4.6 The fundamental insight on how to counteract heterogeneity

It is futile to address the issue of how to sample heterogeneous materials from the point of view that a fixed increment size (one-size-fits-all lot) will be able to deal with whatever heterogeneity features are encountered. It is worse to believe that the correct number of increments/samples *scales* with the total lot mass. These are major mistakes encountered regularly in quite a number of contemporary standards, guidelines and norm-giving documents. It takes the TOS to establish a more realistic understanding.

Any sampling procedure that is not *scaled* with respect to the objectively existing lot *heterogeneity* will not be able to furnish a representative sample. A thorough analysis of these critical issues was published in Esbensen and Wagner[3] and lavishly illustrated in an in-depth source,[4] both of which are recommended as direct follow-up reading to Chapters 3 and 4; Reference 5 is the ultimate textbook on the TOS, which eventually will also be of value to the reader.

4.7 References

1. DS 3077. *Representative sampling—Horizontal Standard*. Danish Standards (2013). http://www.ds.dk, bit.ly/tos2-9
2. L. Petersen, C.K. Dahl and K.H. Esbensen, "Representative mass reduction in sampling—a critical survey of techniques and hardware", *Chemometr. Intell. Lab. Sys.* **74(1),** 95–114 (2004). https://doi.org/10.1016/j.chemolab.2004.03.020, bit.ly/tos4-2
3. K.H. Esbensen and C. Wagner, "Theory of Sampling (TOS) vs Measurement Uncertainty (MU)—a call for integration", *Trends Anal. Chem.* **57,** 93–106 (2014). https://doi.org/10.1016/j.trac.2014.02.007, bit.ly/tos1-6
4. K.H. Esbensen, C. Paoletti and N. Thiex, "Representative sampling for food and feed materials: a critical need for food/feed safety", *J. AOAC Int.* **98(2)** (2015). https://www.ingentaconnect.com/content/aoac/jaoac/2015/00000098/00000002, bit.ly/tos4-4
5. F.F. Pitard, *The Theory of Sampling and Sampling Practice*, 3rd Edn. CRC Press (2019). ISBN: 978-1-138476486

5 "Sampling—is not gambling"

The catchy chapter title is attributed to the founder of the TOS, Pierre Gy. It is a timeless response to the way most practical sampling is still being conducted today. It is an elegant pun, but the statement, of course, needs substantiation. In this chapter, the reader will find simple, easy-to-understand examples of *gambling...* instead of proper *sampling*. It is all about grab sampling and how this approach must be rejected with extreme prejudice. We are on a quest here to make grab sampling extinct! Want to join in?

5.1 Introduction

There are few better examples of an attitude of blind, extreme *hope*, than panning for gold, except perhaps for a Las Vegas roulette wheel! The archetypal gold digger from the time of the great Gold Rush in the Western USA in the latter part of the 19th century illustrates the attitude indicated in Pierre Gy's statement in a romanticised but realistic way.

Unless you were both a gold digger and a qualified geologist with significant insight into the origin of *placer deposits*, and most gold diggers certainly were not, the practice of gold panning was very much extreme gambling.

Most gold diggers were trying out their luck in a specific area, along a specific creek... mainly because it had been *rumoured* that this was a lucky spot, or you ventured out completely on your own luck. Even along the same creek, the likelihood of finding gold nuggets was related to a quite restricted part of the full

Figure 5.1. Gold panning is a wonderful demonstration of "sampling based on hope", a clear form of gambling.

> **Placer deposit:** Natural concentration of heavy minerals caused by the effect of gravity on moving particles. When heavy, stable minerals are freed from their rock matrix by weathering processes, they are typically washed downslope into streams and creeks that quickly winnow out the local bedrock. Thus, the heavy minerals become concentrated in stream, beach and lag (residual) gravel deposits which may not be workable ore deposits in an industrial sense, but which may very well support one-man enterprises. Minerals that form placer deposits have high specific gravity, are chemically resistant to weathering and are durable. Such minerals include gold, platinum, cassiterite, magnetite, chromite, ilmenite, rutile, native copper, zircon, monazite and various gemstones. (Definition modified from Wikipedia)

length only—heavy mineral nuggets travel down a river by saltation along the bottom and find their final resting place precisely where the hydrodynamic force of the water flowing along the bottom layer is no longer able to move the particles further. Even with a geologist's professional knowledge and experience, panning for gold is still a somewhat chancy endeavour. At the time of the gold rush, panning for gold was a very, very low probability gamble, but always with the beguiling potential for the BIG winning gambit lurking just around the next river bend. But obviously, sometimes the gambit does indeed pay off: **gold nuggets!** In fact, one can pan for many other heavy minerals as well which, when concentrated enough, may also achieve favourable results (see box).

But for every winner there were innumerable losers, which is also true of sampling in general if based more on hope than on solid knowledge. Replace unknown potential gold nugget locations along a river bank sediment with an unknown lot heterogeneity: sampling is still not supposed to be gambling!

5.2 Enough analogy

The lessons from the first four chapters comprise a basic conceptual framework, the Theory of Sampling (TOS), necessary to appreciate the crucial role of lot/material *heterogeneity* and a first understanding that all sampling processes *interact* with heterogeneous materials. Sampling is first and foremost directed at *counteracting* the adverse effects of heterogeneity. This chapter is all about how, for every correct, representative sampling performed, there are very many ill-reflected attempts at gambling, with most having even lower odds than the gold panner!

The message here is all about how *grab sampling* does not qualify in the light of heterogeneity. Grab sampling has already been introduced and duly commented upon. All that is needed is a series of practical examples that all have the hallmark of gambling—miles away from proper sampling (Figure 5.2).

Process streams, lots and materials that are significantly heterogeneous are so, both with respect to compositional (CH) and distributional heterogeneity (DH); often but not always, including grain-size segregation as well. All of these are the main enemies in primary sampling stages. A singular, randomly selected extraction of material—meaning a grab sample—from a significantly heterogeneous lot cannot in any way, shape, weight or form hope to catch the characteristics of the *entire* lot, precisely because of DH_{Lot}. This principal understanding, laid out in schematic form in Figure 5.3, is illustrated with examples from the real world of materials in Figures 5.5 and 5.6. There is a very clear impossibility involved in any sampling procedure based on a single-extract operation, grab sampling.

"Well, take a bigger sample, then" is the suggested "remedy" heard most often. It is a completely wrong way of thinking, but it is necessary to take it seriously, in the present didactic context. Figure 5.6 illustrates a wide range of ever larger sample mass *options* (increasing potential sample size), but it is clear that even a wheelbarrow will not necessarily catch a representative sample—it all depends on the magnitude of DH_{Lot}. Even *if* a minor advantage *could* be obtained in the primary sampling stage with such an approach, there is only agony and despair waiting along the subsequent stages in the full "lot-to-analysis" pathway, where extraordinarily large masses are now required to be processed by the analytical laboratory. We are in effect simply "passing the buck" but without having addressed the real heterogeneity-counteracting problem at all.

Pierre Gy's famous statement is a reflection of the irresponsibility involved in hoping to obtain a representative sample without being willing to invest the necessary effort in learning a minimum of the principles in the TOS. Grab sampling is in fact

Figure 5.2. Two manifestations of manual grab sampling—as applied to significantly heterogeneous materials or processes. Alas, this approach is still commonplace in many industrial sectors.

Figure 5.3. Compositional heterogeneity (CH) and distributional heterogeneity (DH). How can a singular, haphazardly selected "sample" (red squares) ever be assumed to be representative of an entire heterogeneous lot?

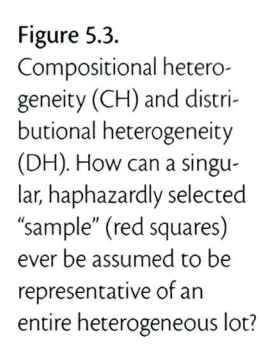

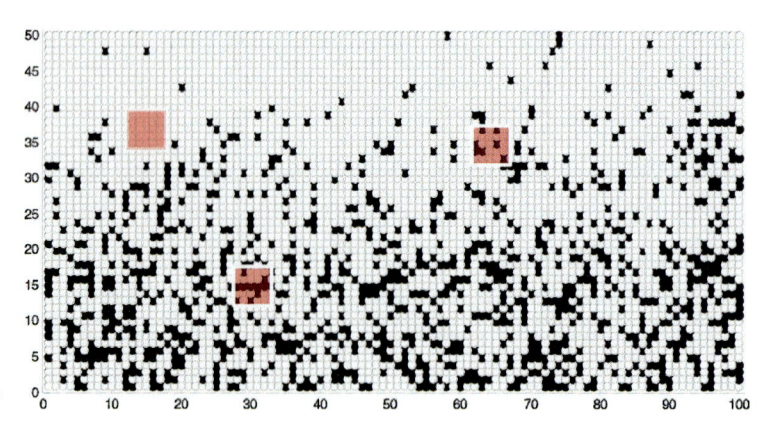

Figure 5.4. Any sampling process interacting with a material that has the characteristics of significant distributional heterogeneity, DH_{lot}, runs a fundamental risk of selecting and extracting worthless, non-representative *specimens*, a far cry from representative samples. The degree of risk is related both to the compositional as well as the distributional heterogeneity in relation to the sample size employed.

Figure 5.5. In addition to CH/DH issues, there may be equally severe grain-size heterogeneity issues for many types of materials (centre panel). The left and right panels show the utter futility of even trying to "cover" this type of heterogeneity, if the chosen (or mandated) sample size is manifestly too small for the job. Often a larger sample mass is *claimed* to be able to solve specific problems, but there is a very narrow limit to any potential benefits from just increasing the proscribed sample mass, see Figure 5.6.

Figure 5.6. What is the correct or increment sample size? No approach of employing "larger samples" will ever be able to counteract heterogeneity effects properly—yet there is a persistent clamour for "bigger samples" in many walks of science, technology and industry. This approach will never eliminate DH, however, unless samples approach the size of the whole lot, obviously an impossible notion. The focus is on the wrong entity—the issue originates with the heterogeneity, not with the voluntary choice of the size of the sampling tool.

just continuing a long-standing tradition of cutting corners in the name of practicality, logistics, work effort and/or the economy of sampling concerning equipment costs, work efforts, personal training etc. As shall become clear below, however, there is a counteracting economic consequence of not getting the sampling representative—that of *hidden economic losses* due to faulty decisions based on faulty analytical results due to non-representative sampling.

So, while a grab sampling *appears* practical, as it always can be carried out with a minimum of effort, and which will, therefore, always end up as the least expensive approach—there is only one counteracting feature, albeit a most serious one: grab sampling can **never** be representative (except by accident)!

Grab sampling amounts to a breach of due diligence. More, in-depth discussion on the merits (*there are none*) and the futility (*unlimited*) of even contemplating grab-sampling can be found all over the TOS literature. Enter *composite sampling* (see Chapters 7 and 8).

6 Pierre Gy's key concept of sampling errors

The founder of the Theory of Sampling (TOS), Pierre Gy (1924–2015) single-handedly developed the TOS from 1950 to 1975 and spent the following 25 years applying it in key industrial sectors (mining, minerals, cement and metals processing). In the course of his career he wrote nine books and gave more than 250 international speeches on all subjects of sampling, including the all-persuasive aspect of hidden economic losses due to neglect of salient sampling issues. In addition to developing the TOS, he also carried out a significant amount of practical R&D. But Gy never worked at a university; he was an independent researcher and a consultant for nearly his entire career—a remarkable scientific life! Gy wrote a five-paper personal scientific testimony which was published in 2004.[1]

Figure 6.1. Pierre Gy (1924–2015), founder of the Theory of Sampling (TOS), at one of his last public appearances (Porsgrunn, Norway, 2005). Photo credit: Kim H. Esbensen.

A comprehensive special issue of *TOS Forum* (https://www.impopen.com/tosf-toc/16_6, bit.ly/tos6-2) is a tribute dedicated to Pierre Gy's life and scientific, technological and industrial achievements.[2] This issue is strongly recommended as a complement to the present chapter.

6.1 Rational understanding of heterogeneity and appropriate sampling

Gy's scientific breakthrough was to take on the overwhelmingly complex phenomenon of *heterogeneity*. The traditional route, taken by most of his predecessors (usually based on various *ad hoc* ideas), was to simplify based on what has later become clear were very unrealistic assumptions.

In his quest to be rational and complete, Gy identified no less than eight sampling errors that represent everything that *can* go wrong in sampling, sub-sampling (sample mass reduction), sample preparation and sample presentation—due to heterogeneity, ill-informed procedures and/or inferior sampling equipment design and performance. Over a period of 25 years he meticulously worked out how to *avoid* committing such errors in the design, manufacture, maintenance and operation of sampling equipment, and elucidated how their adverse impact on the total accumulated uncertainty could be reduced as much as possible when sampling in practice. It was a monumental job. Along the way, he studied for and was awarded two PhDs (in mineral processing and statistics) in order to be adequately equipped to solve the highly complex theoretical and practical problems gradually identified. It is fair to say that, historically, Pierre Gy was the first scientist to tackle the full set of issues related to sampling of heterogeneous materials and processes in a realistic fashion.

As an illustration, consider the few examples of different materials shown in Figure 6.2 (and in Chapters 1–5) and try to imagine what mathematical approach would be appropriate in order to describe their heterogeneity?

Statistics—would very likely be the answer… But what kind, and level, of statistics? Pondering this issue, virtually everyone would likely want to get help via one, or more, type of "statistical distributions", but distributions of what? Analytical results stemming from repeated sampling will not be similar (see earlier chapters), far less identical, but could they follow a distribution of a kind? It may be more-or-less easy to find the right distribution of analytical results, or it may be difficult (very likely the possible answer to this complex issue is intimately related to… heterogeneity). If so, is there one universal distribution ruling over all the

How many sampling errors?
At the first encounter it may be confusing that the number of sampling errors would appear to vary depending upon where and how this issue is treated. Is it 8? Or is it 6? .... It is both! Two sampling errors are only in play when the lot is moving (process sampling), see later. Therefore, there are only six sampling errors in play regarding the present basic introduction to sampling of stationary lots (8 – 2 = 6).

Things to come…
In fact, there is a 9th sampling error, the "*In situ* Nugget Effect" (INE), which shall also be taken into account in more advanced cases, for example related to selection of a given sample or increment mass from unbroken materials; this is especially important in relation to diamond drilling, reverse circulation or blasthole drilling in mining a.o. The INE will be deferred to a planned Book II. Impatient readers will find the INE treated in Reference 11.

Figure 6.2. The manifestations of heterogeneous materials are infinite. What would be the common approach to describing such different types of irregular "distributions" as shown here? And many intrinsic heterogeneity features are not visible to the naked eye, e.g., as illustrated by the truckload of grain shown bottom left. Pierre Gy undertook a 25-year long journey before he reached the destination: the Theory of Sampling (TOS). Photo credit: KHE Consulting (archives), Envirostat, Inc. and iStockphoto

world's very different materials and their very different manifestations? This is definitely where one would like to enlist the experts, the statisticians. Surely this community will know which distribution would be appropriate, and/or will possess the knowledge and competence to find it.

However, one would soon find oneself burdened by the necessity to state to the statisticians exactly what constitute the *physical basis* for the analytical results, the physical *support*. And this would be where the headaches and severe sweating would start to break out, because we are dealing with *heterogeneous*

Figure 6.3. There is no end to the many ways heterogeneity appears. Here a pronounced grain size segregation sometimes expresses itself dramatically (left), at other times completely "embedded" (top-right). It takes a radically different approach to be able to include such features in standard statistical distribution models, in fact Pierre Gy had to invent a new kind of "unit" in order to accomplish this goal, the "heterogeneity contribution". Photo credit: KHE Consulting (archives).

materials. What is the operative basis in the lot for what will become the analytical results?

The physical basis of all analytical results is the analytical aliquot (in the form of a small vial, for example). This view is tantamount to picturing the whole lot as a collection of very many, very small samples, aliquots. However, nobody in their right mind would try to sample a full lot by increments of the size of the miniscule final aliquot. On the other hand, from the approach delineated in earlier chapters, it is clear that all aliquots are the result of a complex, multi-stage sampling and sub-sampling process. And this is the *cardinal knowledge* derived by the TOS: sampling processes can be carried out in many different ways, most of which are demonstrably non-representative. This means that it actually matters which specific sampling process was used to produce the final aliquot. Does this then mean that there are always many, equally possible differing analytical results? Actually yes, **if** one is not acquainted with the TOS and its distinction between representative sampling process, and worthless "specimenting". Before sinking fully into this quicksand, the last thought will likely be… the aliquot **must** be representative of the whole lot: but how to get this vital requirement into the statistics?

Well, the answer to this is much more complex than might be suggested by a first reflection. First of all, the statisticians do **not** know the world's very different materials and their very different heterogeneities; why should they? It is **you**, the sampler, who is the expert on materials here. After some more reflection based on your own experiences, it will become clear that whereas statistics is addressing a population of "units", each of which are *identical* except for the differing analytical results, the "units" of heterogeneous materials and lots are defined by the specific nature of the material **and** the specific sampling procedure used, in particular by the increment, or sample, size (mass). Thus, in a very real sense, the final analytical results depend materially on the sampling procedure—a different sampling approach, e.g. grab sampling vs composite sampling, will assuredly give rise to different analytical results (with the composite sampling result being overwhelmingly more reliable). Go tell this to the statisticians! A direct inroad to this complexity, without all the math, can be found in Reference 3.

At the outset, then, we are not in a position to simply take over conventional statistical notions, populations, units. We are left to fend for ourselves: and this was exactly what Pierre Gy realised, and which led him to the ambitious goal of developing

And it gets worse
Add to this that this sampling process ambiguity also characterises every sub-sampling stage involved before producing the final analytical aliquot. The TOS is needed to guide all sampling stages, no exception.

👆 bit.ly/tos1-5

the "appropriate statistics" with which to be able to describe the real-world of heterogeneous materials (and processes). This is indeed more complex than a world conceptualised via scalars, conventional vectors, matrices and other arrays of data, which lend themselves to simple descriptions by statistical moments of distributions: averages, standard deviations, variances etc.

> Pierre Gy's new lot "unit" was the *heterogeneity contribution*. Each unit of a lot, be this fragments or increments, can be thought of as carrying a specific "heterogeneity contribution" with respect to the heterogeneity of the total lot, here realised as increments pertaining to a specific sampling situation. The heterogeneity contribution is a simple mass-weighted relative compositional deviation with respect to the average lot grade (the average process grade in this example), see Chapters 3 and 4. Although extremely simple, this *compound* measure was exactly what was needed in order to break up the theoretical gridlock Gy had been struggling with for many years. The heterogeneity contribution actually allows both the compositional heterogeneity of each unit (CH_i) and this unit's mass to be taken into account simultaneously. It turned out that conventional statistical description of the distribution of the heterogeneity contributions, **not** of the analytical results directly, was the defining breakthrough.

For the interested reader, here are several, carefully crafted next-level introductions to this journey.[4–7]

6.2 Although complex, there is hope

So, although complex, the TOS can in fact be made easily accessible. There are many systematic elements in the TOS, which makes mastery possible, even from a less in-depth theoretical and mathematical level.

For example, the TOS' six basic sampling errors relate to three sources only: the *material* (always heterogeneous, it is only a matter of degree), the *sampling equipment* (which can be designed either to promote a representative extraction, or not) and the *sampling process* (even correctly designed equipment can be used in a non-representative manner). In general, sampling is also defined by whether the lot is *stationary* or *moving* when sampling takes place; a distinction that is well-known within many application fields. For example, in the realm of powders, "sample only when the powder is moving" is an empirical experience paralleling one of the TOS Governing Principles (GP), but generalised to the principle of Lot Dimensionality Transformation (LDT).

This was a breakthrough: as it turned out, *some* sampling errors were found to be able to be *eliminated* completely, which simplifies the sampling agenda considerably. But it is, of course, necessary to know **how** to identify these errors competently and exactly **how** to eliminate them. These are the so-called Incorrect Sampling Errors (ISE), a concept that was to be instrumental in order to be able to give a distinct definition of both a "correct" as well as a "representative" sampling process.

Following this rational simplification route, recently the TOS has been presented in a fully axiomatic framework.[8–10] Figure 6.4 shows the systematics of the complete framework of TOS' General Principles (GP), Sampling Unit Operations (SUO) and the rules for managing all eight sampling errors, distinguished as ISE and Correct Sampling Errors (CSE). The value of this overview is that all principal elements needed for a guaranteed representative pathway "from-lot-to-aliquot" are outlined here. This framework should be viewed as an enabler for delving into the TOS literature more fully without the danger of losing one's way among so many, more-or-less new concepts and relationships. The reader is encouraged to refer to this diagramatic TOS

Sampling error terminology
Sometimes the acronym ISE means "Increment ... Error", but sometimes it is read as "Incorrect ... Error". This slight confusion has a historical root, but it does not matter once this duality has been understood; what matters is that all ISEs, of whatever denomination, are incorrect.
IDE: Increment Delimitation Error
IEE: Increment Extraction Error
IPE: Increment Preparation Error
IWE: Increment Weighing Error

(IEE was earlier known as IME: Increment Materialisation Error)

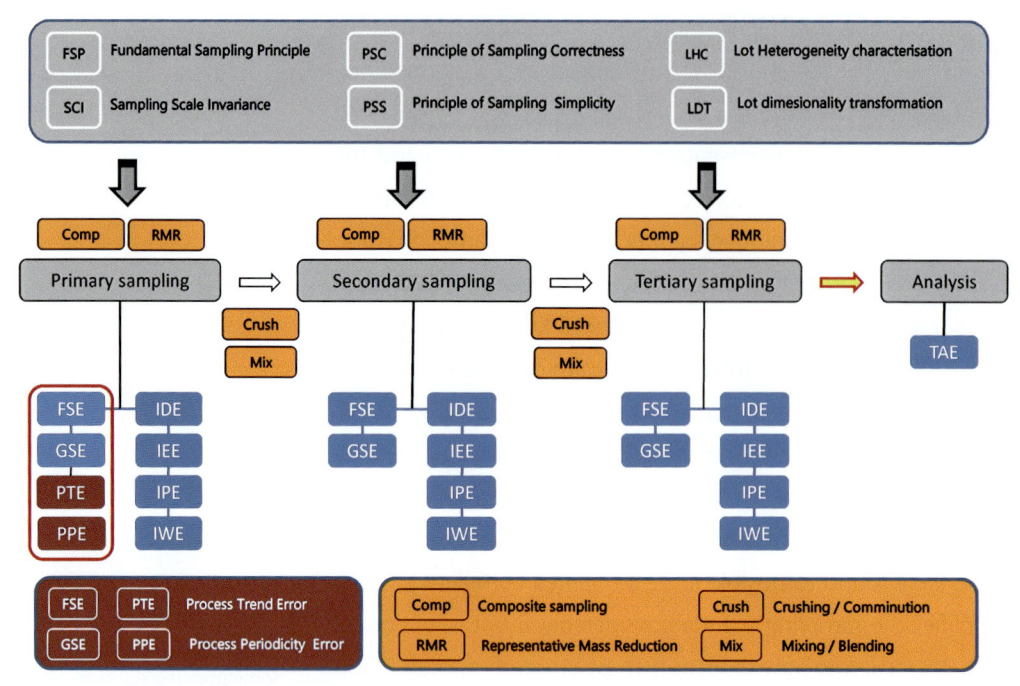

Figure 6.4. Theory of Sampling (TOS) synoptic overview, shown here for three sampling stages. TOS is comprised by six Governing Principles (GP) (top grey panel), four Sampling Unit Operations (bottom yellow panel) and eight sampling errors (blue/maroon). The basic differentiation between the ISEs (IDE, IEE, IPE, IWE) and the CSEs (FSE, GSE) is treated in the text where appropriate. For details, see References 11–13 and other authoritative references therein. Illustration copyright KHE Consulting; reproduced with permission.

overview whenever needed as further new aspects of the TOS are introduced.

6.3 How to sample representatively: the TOS

The first task on **any** sampling agenda is to eliminate the ISEs, mainly an issue regarding the design, installation, operation and maintenance of the sampling equipment. Subsequently, what remains, are the CSEs which can be dealt with by more standard means, i.e. by increasing the number of composite sampling increments, Q, with respect to the empirical heterogeneity encountered—but always honouring the Fundamental Sampling Principle, FSP and always involving the pertinent other GPs and SUOs where and when needed. Recent general introductions to the TOS are References 8, 9, 13 and 14. In relation to the specific disciplines of chemometrics and multivariate data analysis, as well as in relation to pharmaceutical production for example, dedicated introductions can be found in References 5 and 15, respectively.

Thus, Figure 6.4 depicts a generic, multi-stage sampling process framework, the singular purpose of which is to deliver a representative analytical aliquot (represented by the yellow arrow to the right). Sampling of stationary lots makes use of six basic sampling errors (blue), while process sampling (sampling of dynamic lots) needs two more errors to be tackled in full (maroon).

The TOS logically demands that all pre-aliquot steps are supervised and governed by a *unified* sampling responsibility. This is a *legal person*, either in the form of a single individual (a "sampling czar") or by a committee representing all departments in which sampling is performed. The latter situation is typical of very many organisational solutions in big companies and corporations; but is unfortunately also the reason behind a considerable proportion of the sampling problems met with in real life. Experience with many big corporations, companies and organisations unfortunately points to considerable difficulties clearly only related to inter-departmental strife or lack of proper collaboration. There are many cases where traditional rivalries between departments, individuals or just historical traditions make effective sampling across the entire "lot-to-analysis" pathway well-nigh impossible.[a] For such

---

[a] This is the situation in more than 75% of cases in which the author of this book has been involved! Which is not to say that the technical elements of the TOS lose their importance—but such organisational dysfunction of course makes it more difficult to apply the TOS' powers effectively.

cases, solutions rather come from the realm of organisational psychology.

However, disruptive solution possibilities also exist within the realm of the TOS: it is most certainly not only the front-line samplers who need proper education with respect to the TOS. Of course, such individuals are of primary importance because they collect the samples upon which analytical results ultimately are produced and upon which important decisions are made. Indeed, all individuals who are responsible for optimising sampling and process performance should feel a responsibility to become conversant with the TOS. Thus, whether company president, vice president, operations manager, process technician, laboratory supervisor, quality assurance and quality control manager and indeed also concerned investors and company shareholders—all need a succinct understanding of the basic principles in TOS in order to be fully competent in their respective roles and capacities, as this will certainly affect the economic bottom line (hidden costs galore).

Thus, it is all about how to be able to identify and eliminate, or reduce, a small group of sampling errors—this is one of the greatest legacies of Pierre Gy.

The reader is recommended to consult References 4, 7 and 8 immediately after reading this chapter.

6.4   References

1. P. Gy, "Sampling of discrete materials—a new introduction to the theory of sampling I. Qualitative approach", *Chemometr. Intell. Lab. Syst.* **74**, 7–24 (2004) https://doi.org/10.1016/j.chemolab.2004.05.012, bit.ly/tos6-1a; P. Gy, "Sampling of discrete materials II. Quantitative approach—sampling of zero-dimensional objects", *Chemometr. Intell. Lab. Syst.* **74**, 25–38 (2004) https://doi.org/10.1016/j.chemolab.2004.05.015,

👆 bit.ly/tos6-1b; P. Gy, "Sampling of discrete materials III. Quantitative approach—sampling of one-dimensional objects", *Chemometr. Intell. Lab. Syst.* **74,** 39–47 (2004) https://doi.org/10.1016/j.chemolab.2004.05.011, 👆 bit.ly/tos6-1c; P. Gy, "Part IV: 50 years of sampling theory—a personal history", *Chemometr. Intell. Lab. Syst.* **74,** 49–60 (2004) https://doi.org/10.1016/j.chemolab.2004.05.014, 👆 bit.ly/tos6-1d; P. Gy, "Part V: Annotated literature compilation of Pierre Gy", *Chemometr. Intell. Lab. Syst.* **74,** 61–70 (2004) https://doi.org/10.1016/j.chemolab.2004.05.010, 👆 bit.ly/tos6-1e

2. "Pierre Gy (1924–2015)—in memoriam", *TOS Forum* Issue 6 (2016). https://www.impopen.com/tosf-toc/16_6, 👆 bit.ly/tos6-2

3. K.H. Esbensen, "Materials properties: heterogeneity and appropriate sampling modes", *J. AOAC Int.* **98,** 269–274 (2015). https://doi.org/10.5740/jaoacint.14-234, 👆 bit.ly/tos1-5

4. K.H. Esbensen, "Sampling – theory and practice", *Alchemist* **Issue 85,** 3–6 (August 2017), London Bullion Market Association. https://kheconsult.com/wp-content/uploads/2017/11/Alch85Complete-1.pdf, 👆 bit.ly/tos6-4

5. K.H. Esbensen, R.J. Romanach and A.D. Roman-Ospino, "Theory of Sampling (TOS) – a necessary and sufficient guarantee for reliable multivariate data analysis in pharmaceutical manufacturing", in *Multivariate Analysis in Pharmaceutical Industry*, Ed by A.P. Ferreira, J.C. Menezes and M. Tobin. Academic Press, Ch. 4 (2018). https://doi.org/10.1016/B978-0-12-811065-2.00005-9, 👆 bit.ly/tos6-5

6. K.H. Esbensen and P. Paasch-Mortensen, "Process sampling: Theory of Sampling – the missing link in Process Analytical Technology (PAT)", in *Process Analytical Technology*, 2nd Edn, Ed by K.A. Bakeev. Wiley, Ch. 3 (2010). https://doi.org/10.1002/9780470689592.ch3, 👆 bit.ly/tos6-6

7. J. AOAC Int. **98(2)** (2015). https://www.ingentaconnect.com/content/aoac/jaoac/2015/00000098/00000002, 👆 bit.ly/tos2-5
8. K.H. Esbensen and C. Wagner, "Why we need the Theory of Sampling", *The Analytical Scientist* **Issue 21**, 30–38 (2014).
9. K.H. Esbensen and C. Wagner, "Theory of Sampling (TOS) versus measurement uncertainty (MU) – a call for integration", *Trends Anal. Chem.* **57**, 93–106 (2014). https://doi.org/10.1016/j.trac.2014.02.007, 👆 bit.ly/tos1-6
10. DS 3077. *Representative Sampling—Horizontal Standard*. Danish Standards (2013). www.ds.dk, 👆 bit.ly/tos1-2
11. F.F. Pitard, *The Theory of Sampling and Sampling Practice*, 3rd Edn. CRC Press (2019). ISBN: 978-1-138476486
12. P. Gy, *Sampling for Analytical Purposes*, 1st Edn. Wiley, New York (1998). ISBN: 978-0-471-97956-2
13. R.C.A. Minnitt and K.H. Esbensen, "Pierre Gy's development of the Theory of Sampling: a retrospective summary with a didactic tutorial on quantitative sampling of one-dimensional lots", *TOS Forum* **Issue 7**, 7–19 (2017). https://doi.org/10.1255/tosf.96, 👆 bit.ly/tos6-13
14. K.H. Esbensen and L.P. Julius, "Representative sampling, data quality, validation – a necessary trinity in chemometrics", in *Comprehensive Chemometrics*, Ed by S. Brown, R. Tauler and R. Walczak. Elsevier, Oxford, Vol. 4, pp. 1–20 (2009). https://doi.org/10.1016/B978-044452701-1.00088-0, 👆 bit.ly/tos1-3
15. K.H. Esbensen and B. Swarbrick, *Multivariate Date Analysis – An Introduction to Multivariate Data Analysis, Process Analytical Technology and Quality by Design*, 6th Edn. CAMO Software AS (2018). ISBN 978-82-691104-0-1, https://www.amazon.com/Multivariate-Data-Analysis-introduction-Analytical/dp/826911040X/, 👆 http://bit.ly/chemometrics
16. K.H. Esbensen and C.A. Ramsey, "QC of sampling processes—a first overview: from field to test portion", *J. AOAC. Int.*

98, 282–287 (2015). https://doi.org/10.5740/jaoacint.14-288, 👆 bit.ly/tos6-16

7 Composite sampling I: the Fundamental Sampling Principle

The first chapters have presented many examples of heterogeneous lots and their varied manifestations—and stressed the resulting difficulties when less appropriate sampling procedures are employed, especially grab sampling. Now is the time to start addressing the reasonable question: what can be done about all this adverse heterogeneity? Luckily, there are many actions available that will help, all stemming from the Theory of Sampling (TOS). First to be introduced is the powerful concept of *composite sampling* in close relation with the Fundamental Sampling Principle (FSP). These are, in fact, the only two options available at the *primary sampling* stage of any lot, i.e. when facing the original sampling target and are, therefore, of paramount importance for all sampling, at all scales, for all materials…

AA
FSP: Fundamental Sampling Principle
FSE: Fundamental Sampling Error
GSE: Grouping and Segmentation Error
SUO: Sampling Unit Operations
GP: Governing Principles
RE: Replication Experiment
BBS/TSR: Bed-Blending Stacking/Thin-Slice Reclaiming
LDT: Lot Dimensionality Transformation

7.1 WHAT TO DO with all this heterogeneity?

Recapitulation: Trying to sample any heterogeneous lot with a single sampling operation is completely out of the question, for the simple reason that a single sample (*"specimen"* rather) will never be able to *represent* the heterogeneous material (lot) all by itself, except in the rarest of accidental situations (and one will never be able to know **if** nor **when** this was the case). This is regardless of whether the heterogeneity is visible or not. This latter point is worth emphasising because of the frequent situation of apparently visible uniformity, see Figure 7.1.

Figure 7.1. A single grab sample (specimen) is never able to represent the entire lot because it is manifestly not able to *cover* any material with a significant compositional heterogeneity and distributional heterogeneity. This important truth holds for lots of all sizes, shapes and forms.

Figure 7.2. A set of individual grab samples (Q increments) destined to be aggregated goes a long way towards *covering* the lot in question, but certain further demands must also be met. It is not only about the numerical magnitude of Q, i.e. it is not only about *how many* increments are used, but it is just as much about *how* these are *deployed* geometrically, i.e. are they covering the entire lot *volume*, or not. This illustration is also pertinent to the issue of introducing a number of TOS' sampling errors, in this case fundamental sampling error (FSE) and grouping and segmentation error (GSE), see also Figure 3.5.

But there is hope—indeed an obvious solution is immediately available. While in Figure 7.2 each individual grab sample (white circles) will fail for this reason—there is much more of a constructive chance for an *ensemble* of such: a *composite sample* is defined as an *ensemble* of individual **increments**, carefully spread out so as to *cover* the full geometry of the lot in question, with the express intention to be accumulated into one aggregate, a composite sample. The notion of a *composite sample*, subject to a few, logical demands, will be shown to be the saviour of pretty much all sampling problems and issues that otherwise would have been left unsolvable. *Composite sampling* constitutes one of four sampling unit operations (SUO) to be described more fully below; another SUO, *mixing*, was described in Chapter 4.

For composite samples, the number of increments (Q) is critically important, but **only** when deployed in a problem-dependent fashion, indeed **only** when *covering* the full lot volume adequately. Rather paradoxically compared to conventional statistics, it is not *only* the number of samples/increments that is important, but specifically also *how* these Q increments are *deployed*. For the situation depicted in Figure 7.2, a reasonable sampling coverage is beginning to be achieved as Q increases... This is a situation easily depicted for a 2-D lot, where the third dimension is covered if/when all increments cover the same depth interval.

However, the situation becomes significantly more complex regarding stationary 3-D lots, e.g. piles, silos, vessels etc. In Figure 7.3, it is obviously not a solution to deploy Q increments only within a local, narrow footprint—this gets nowhere, not even trying to cover the entire lot (left panel). What is needed is to broaden out the sampling plan, but most emphatically not only along the lot surface. It is imperative that the coverage is also able to sample all potential increment locations in the *interior* of the lot.

Figure 3.5. Significantly heterogeneous material, as sampled wrongly (A) and with a much better chance of being representative (B). Note that the "overkill" 35-increment example is only used to emphasise the point that composite sampling needs to make use of the optimal, i.e. the necessary and sufficient number of increments, Q, with which to counteract the specific heterogeneity met with. Finding Q is one of the two major objectives of representative sampling.

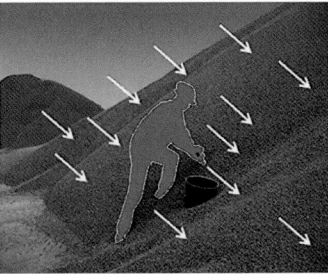

Figure 7.3. Local vs broad deployment of the same number of increments, Q. Note, however, that while the broadened-out deployment plan in the right panel does a much better job of "covering" the surface of the lot, it does not sample from within the significantly larger inner volume? How to "cover" a full-fledged 3-D lot? Dig in!

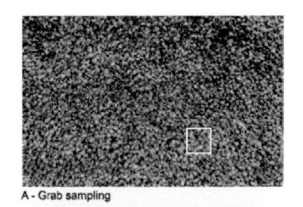

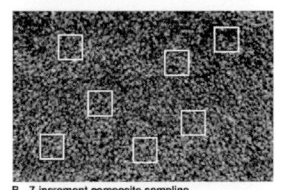

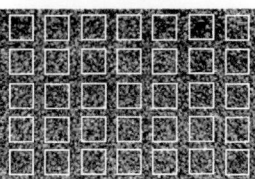

Enter the **Fundamental Sampling Principle (FSP)**: "All *virtual* increments in any lot must be available for sampling and must have the same probability of ending up in the final composite sample… All *potential* increments that might be identified in a composite sampling plan *must* be available for practical physical extraction, no exception".

This means that even the sampling taking place in the right panel in Figure 7.3 cannot be said to uphold the FSP!

Composite sampling is necessary, but in itself not a panacea; it must always comply with the demands of the FSP as well. Thus composite sampling is a *necessary* condition, but only becomes *sufficient* when also obeying the FSP. The FSP constitutes the first of six Governing Principles (GP) in the TOS. These principles express the imperative demands that must apply identically to lots of all dimensionalities, 0-D, 1-D, 2-D as well as 3-D lots.

Figures 7.4–7.7 illustrate various important aspects of the necessary compliance between composite sampling and the FSP.

1-D lots are not really 1-D lots like a geometric line, but lots in which one of the three dimensions completely *dominates* the two others, Figure 7.6. While being 3-D lots in principle, because of the TOS' demands that any increment extracted from such a lot must cover the two other dimensions completely, this lot actually becomes a 1-D lot in practice along the remaining extension dimension, because it is now only the heterogeneity in the elongated dimension that matters—all "transverse heterogeneity" has been successfully represented in each cross-stream slice (increment).

Figure 7.6 shows a powder manifestation of a 1-D lot (the lot material is power plant incineration ash which needs characterisation and hence primary samples are collected). But, in fact, Figure 7.6 illustrates the procedure used in the laboratory for sub-sampling the primary samples. Here compliance with the FSP is secured through application of the operation of

Composite Sampling I: the Fundamental Sampling Principle

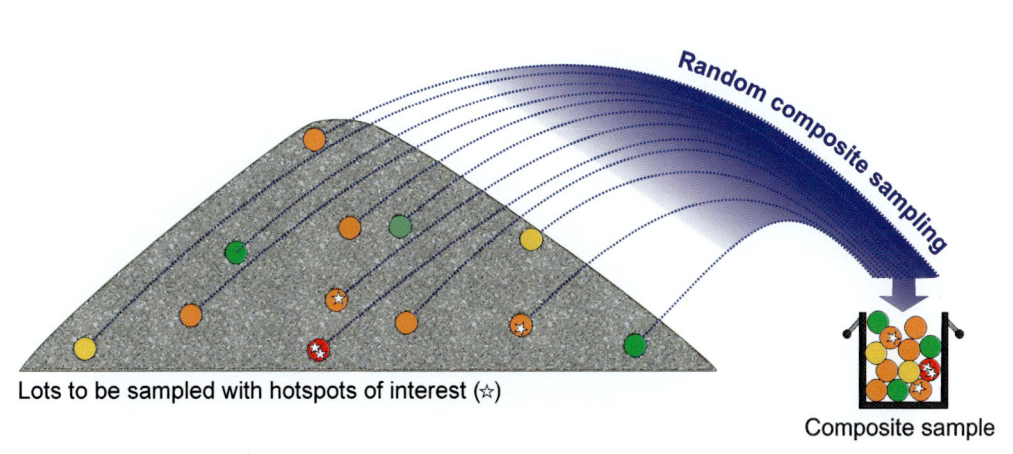

Figure 7.4. Principal illustration of composite sampling complying with the demand of full lot volume coverage (FSP). Based on this, the only requirement left is using an appropriate number of increments, Q, sufficient for fit-for-purpose representativity; for example as assessed by a Replication Experiment (RE) or a more involved approach. Illustration courtesy Mr Martin Lischka, with permission.

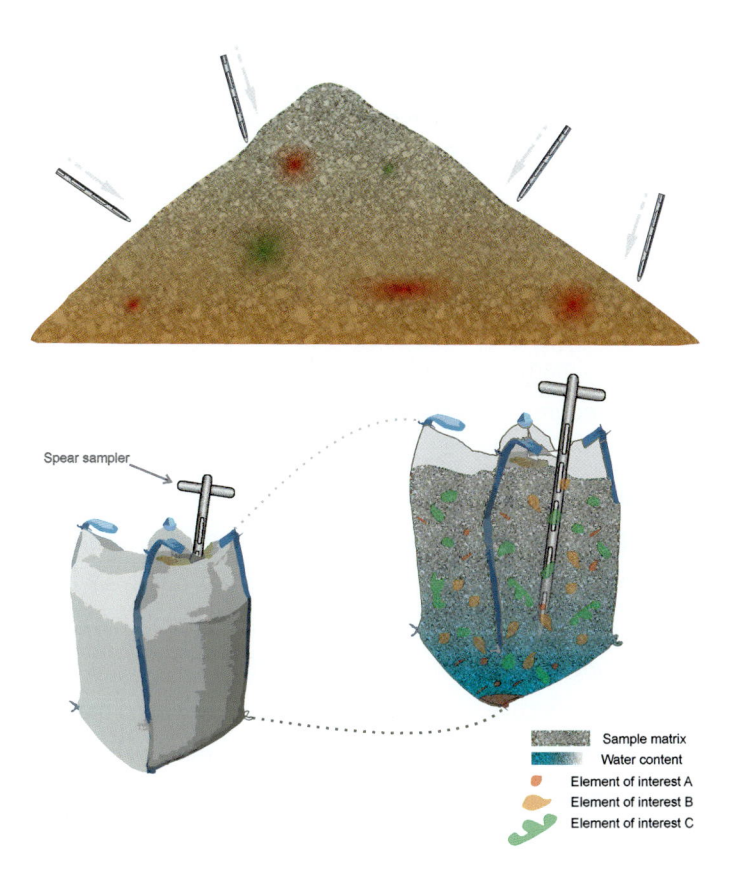

Figure 7.5. The FSP demands that all potential increments must have the same, non-zero, probability of being extracted. It is emphatically not enough to broaden out a sampling plan only along the *surface* of a 3-D lot. The inner volume (by far the largest volume fraction of any 3-D body) must be available for sampling with the same accessibility and ease. Note that in addition there may be special issues also placing demands on the sampling procedure, e.g. free moisture displaying a distinct tendency towards segregation. All ISE + CSE manifestations must be eliminated or reduced. Illustration courtesy Mr Martin Lischka, with permission.

Figure 7.6. Bed-blending technique applying composite sampling and full lot coverage. The primary sample material is first laid out in a multiple-layer stacking operation, in this case in six layers (called "lines"). In the literature this technique is commonly referred to as bed-blending stacking, a particularly efficient form of preparing for mixing which takes place by the subsequent slicing of increments. Note how the thin extraction device covers the full width and depth of the stacked material and thus covers the two transverse dimensions completely as required. The lower panel shows six out of a total of seven such transverse thin-slice extractions. This ingenious two-step procedure allows complete access to the original lot volume, thus in full compliance with the FSP. This procedure can, in fact, be applied to all lot sizes from laboratory sub-sampling to bulk materials handling.

"bed-blending". *All* of the primary sample material is laid out in the sampling rack—in this particular case in six layers, but preferentially as many as possible, after which the cross-cutting sampling procedure enjoys complete access to "everywhere" in the lot.

In this example, seven such transverse increments were extracted, but which in reality corresponds to no less than 42 increments in total, since the material was laid out in six beds originally. This compound composite sampling approach is called *bed-blending stacking/thin-slice reclaiming* (BBS/TSR). This is obviously a very effective approach when it is acknowledged that the total number of increments scales with the number of beds laid down (B) × the number of full transverse cuts extracted (Q), in the above 6 × 7 = 42. It is worth noting that each reclaiming thin-slice

Composite Sampling I: the Fundamental Sampling Principle

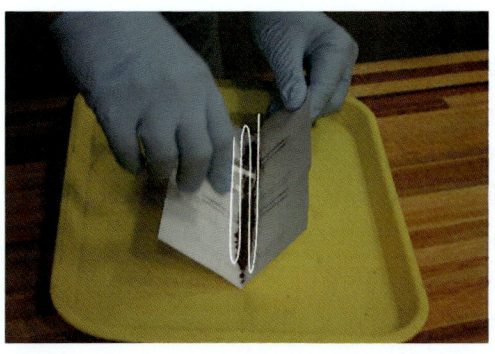

Figure 7.7. Even with "difficult materials" (coarser grains, "clumpy constitution"), bed-blending/thin-slice reclaiming is still often possible. Here an example of a particularly inexpensive laboratory improvement is demonstrated, essentially at no cost. A willingness to invest just a little effort to understand the principles of the TOS is all that counts. N.B. the sampling procedure developed in this example ended up using more increments than those indicated here... in fact the laboratory involved developed several versions (each with its own specific Q targeting the specific heterogeneous materials met with).

increment is in effect a small B-composite sample in its own right, made up of B constituting layer-increments. These $B \times Q$ increments demonstrably cover the entire geometric volume of the precursor lot very well, irrespective of its original form, geometry and mass. This is because one has made the effort to string out the complete lot material in a 1-D linear manifestation (in a folded format), making compliance with FSP both easy and *very* effective. The combined operation is a kind of *Lot Dimensionality Transformation (LDT)*, in this case from 3-D to 1-D.

Note that this technique can be applied to any scale and is, in fact, often used for primary lot sampling and blending/mixing purposes of material occurring in significant tonnages as well (bulk materials handling).

By extracting several increments at regular intervals along the elongated dimension (or at random positions), a particularly effective sampling is achieved by aggregating all increments. By this approach the entire lot volume (the entire primary sample volume) is *guaranteed* to be available for sampling and this

In fact the BBS/TSR approach combines LDT and composite sampling with extremely effective mixing.

composite sampling process therefore complies entirely with FSP. This is of interest also for coarser fragment aggregates, which traditionally are considered as "difficult" to sample.

Figure 7.7 shows such a case subjected to "bed-blending stacking/thin-slice reclaiming" in an impromptu implementation. Note that this technique is not necessarily associated with a particular type, or brand, of equipment—on the contrary: until this type of laboratory sampling was demonstrated (for a world-class company with a strong laboratory division), simple spatula-based grab sampling had been ruling for years/decades… "because there is no other equipment available". Well now there is: the BBS/TSR *procedure* can be implemented in many fashions; specific equipment for this ingenious approach is currently under development—but you may also, if you so desire "make your own".

Which materials, which company, which laboratory in the above illustrations is of absolutely no interest—the only thing that matters is that a simple, essentially no-cost solution [a piece of cardboard (folded) and a high-walled spatula] was able to transform the world's worst sampling procedure (grab sampling) to an unsurpassed, representative procedure (bed-blending stacking/thin-slice reclaiming) because of understanding and respect for the TOS in general, and for the FSP in particular. Figures 7.7 and 7.8 are role model examples of sampling procedure improvements.[a]

There are many other variations on the theme of composite sampling + FSP in the world of science, technology and industry,

---

[a]*Some* readers may perhaps take issue with the technical quality of Figure 7.7, which is readily acknowledged by the author. But such a complaint fades into insignificance because of the superior real-world relevance of this snapshot documentation of a major laboratory breakthrough… After this was acknowledged, came the inspired readiness to implement this new procedure making good use of the in-house company machine shop.

Composite Sampling I: the Fundamental Sampling Principle

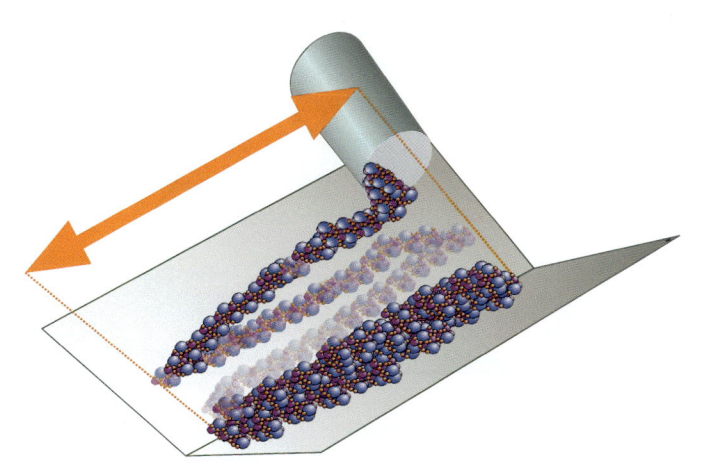

Figure 7.8. Universal principle of Bed-Blending Stacking, applicable to all scales, as the necessary preparation for "Thin-Slice Reclaiming" (BBS/TSR), see also Figures 7.6 and 7.7. Illustration courtesy Mr Martin Lischka, with permission.

but the present introduction should allow one to recognise these easily. The next chapter presents more examples of the versatility and effectiveness of composite sampling, especially for sampling dynamic lots, i.e. moving streams of matter (process sampling).

> **One finds many things on the internet... but are they true 1?**
>
> This is, of course, a very broad question, but here it means: "One finds many things about *sampling* on the internet, but are they true, specifically are they representative?". There is no better way to appreciate the message in this box than to start by consulting the source itself, in this case a YouTube (2 min) video: https://www.youtube.com/watch?v=2qxQ6M4cq8w, 👆 bit.ly/tos7-1. Pay close attention to the statement at 1:20–1:24 seconds in the video clip, and compare with your growing TOS knowledge!
>
> This video purports to teach the viewer how to take a representative sample from an asphalt delivery truck. It is a good video, crisp, superior realism, good accounting of, for example, safety boots, gloves, helmet, the sampling workflow etc. There is really only one thing wrong with it: the sampling method portrayed and recommended is structurally flawed and will sadly **never** be able to lead to a representative sample of the truck load.

Here is why:

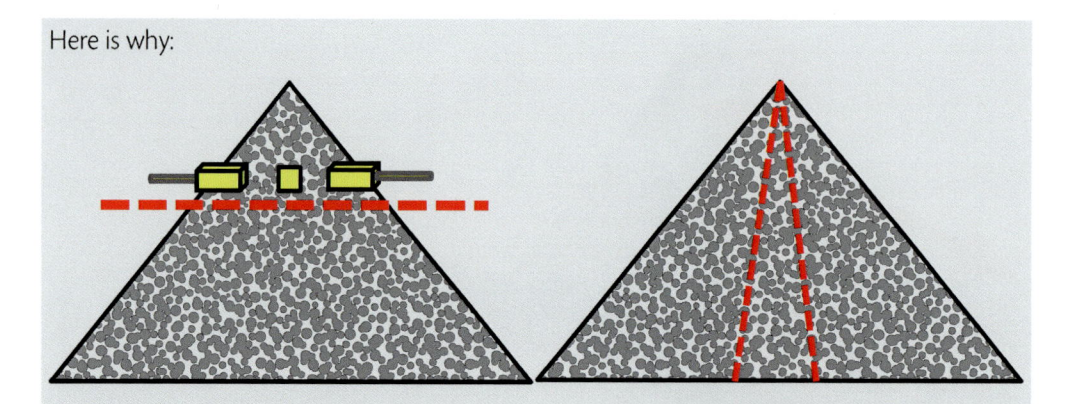

With the instructions given in the video, it will never be possible to collect increments below the red line: which means that more than 80 % of the volume of the truck load will never be sampled. There is consequently no way that the four-increment composite sample approach shown in the video can ever be "representative" of the whole truck load lot. With TOS expertise on board, you should easily be able to state, with confidence: "This approach is in blatant breach of the Fundamental Sampling Principle (FSP)".

It is instructional to ponder *how* this "four-increment-top-20 %" approach could have been misconstrued as representative? This is no doubt caused by the tacit assumption that is met with very, very often but almost never articulated: "This material is likely so *uniform* in composition (did anyone hear "homogeneous"?) that just about any grab sample will do! While the demonstrated approach makes a serious effort to compensate for peripheral heterogeneity (four increments, 90° apart), there is absolutely no similar concern regarding the possible vertical heterogeneity on the truck load lot. The right-hand illustration shows the only correct increment delineation, a far cry from what is recommended!

One finds many things on the internet... but are they true ?

While you are at, it you may test yourself with another instructional "Asphalt sampling and testing" video: https://www.youtube.com/watch?v=T374R-0UfLw, 👆 bit.ly/tos7-2

Here one finds a very good overview of the many routine, as well as advanced testing methods employed by the asphalt industry. From a testing point of view this video is comprehensive and compelling. The video itself ends with the claim that: "...make asphalt testing a truly scientific endeavour". This is indeed true for testing and measuring on samples brought to, or which have been sub-divided in, the professional analytical laboratory. However, reiterating the very first sentence in this book, "What is the meaning of analysing a non-representative sample? None—it is pointless!", now please view this video a second time, this time focusing only on identifying the crucial primary sampling stages shown. There are two instances, in one of which the commentary emphasises that "a representative sample must be collected". Pay close attention to these two instances: do the primary sampling approaches demonstrated *comply* with the principles and rules laid down by the TOS? If not, what is wrong?

Key lessons to be learned: There is much talk about representative samples on the internet, but is this crucial characteristic ever properly *defined*? Many talk the talk, but do they walk the walk? From the principles and rules presented in this book it is easy to demonstrate whether a specific sampling procedure is representative, or **not**—the TOS to the fore!

Finally, test your well-earned TOS skills one last time by careful scrutiny of a third video; sooner or later one is bound to find close to full TOS compliance and this is a nice example: https://www.youtube.com/watch?v=hNscKsLLQbI, ☝ bit.ly/tos7-3

8 Composite sampling II: lot dimensionality transformation

The lesson so far is very clear: grab sampling never works—instead composite sampling rules! In the previous chapters we met with many examples of lots made up of different materials with widely different heterogeneities, lots at all scales. For all lots the Fundamental Sampling Principle (FSP) forces the active sampler to "cover" the entire lot volume meticulously with an appropriate number of increments (Q). While easy to understand in principle, it must be acknowledged that the practicality of reaching every corner of every possible lot can be a daunting task, especially if the lot is a three-dimensional (3-D) body much larger than what can easily be manipulated on the laboratory bench. At the primary sampling stage, the sampler meets with all manner of lot sizes from handy to large (to enormous) from which to take a primary sample, a composite sample. Is there an easier way to do the right thing, easier that just to "dig in"? Luckily there is! First, we will appreciate how easy it is to carry out composite sampling on a one-dimensional (1-D) lot. And the whole world will suddenly get a lot easier—in fact we will be able fully to appreciate just how powerful representative sampling can be!

AA
FSP: Fundamental Sampling Principle
PSD: Particle Size Distribution
LDT: Lot Dimensionality Transformation

8.1 1-D lots: *conveniently* elongated lots

The principal characteristic of all 1-D lots is that the transverse dimensions (width, thickness) are so small (and reasonably constant in absolute magnitude) that all increments will be able to "cover" these two dimensions *fully*. This is the essential point. Unfortunately, the real world does not always comply easily though. Figures 8.1 and 8.2 illustrate this issue.

Figure 8.1. A: Archetypal moving 1-D lots, the pipeline (left) and the conveyor belt (right), with potential sampling station locations indicated by red arrows. B: The folly of grab sampling in the process domain. Manual process grab sampling attempts to cover the full transverse dimensions of the flow (width/height) but with clear dangers of being insufficient—grossly so in two of the examples shown. A sampling bias has been introduced, which will haunt the reliability all the way to analysis. More on the sampling bias in several later chapters where appropriate.

Figure 8.2. Schematic illustration. Increments can be extracted from 1-D lots in a variety of ways, but only one is correct (representative): a complete *slice* of the stream/flow defined by *parallel boundaries* (shown in yellow). All other increment delineations shown are incorrect and will give rise to a sampling bias. For process sampling, e.g. a pipeline, the correct increment delineation is also a planar–parallel slice of the flow, i.e. a cylindrical cut (shown in grey).

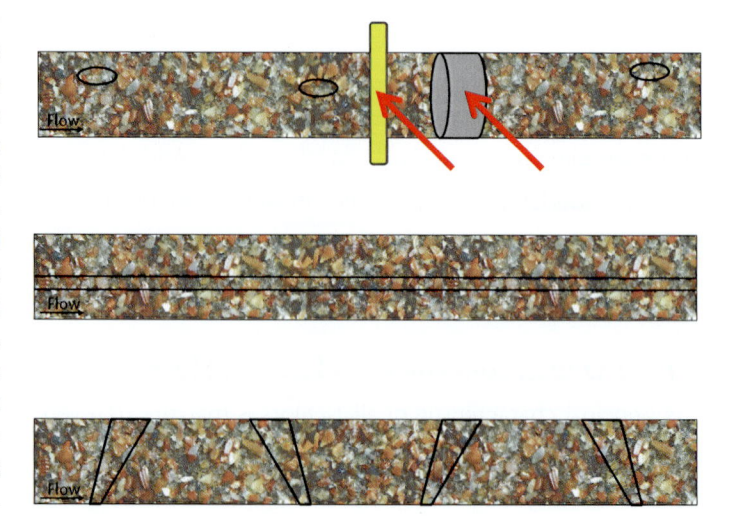

A 1-D lot can either be a stationary lot or it may be a dynamic, i.e. a moving, lot. The latter is the archetypal "moving stream of matter", which may be made up of material systems with many phases each with their own particle size distribution (PSD). For the present illustration we focus on aggregate streams of matter (could also be slurries)—the arguments to be presented are valid for all types of compositional systems *on the move* or strung out as a stationary 1-D lot.

Figure 8.2 shows how the adverse sampling bias can be produced by inappropriate sampling. But there is hope; especially if the sampler invests just a trifling effort, based on the TOS' principles.

8.2   Process sampling

Figure 8.3 gives two examples of 3-D lots that, at some point in their life, actually manifest themselves as moving 1-D lots. The fish elevator example is explained in the figure caption; the grain truck example is but the terminal end of a much bigger off-loading process from a 30,000-ton cargo ship carrying feed soy beans. The entire nine-cargo-hold soy bean lot is off-loaded by one-ton crane grabs being deposited in the hopper shown. Each truckload, 10 tons, is then driven to the warehouse shown in the background, where the complex 3-D ship cargo is now again turned into a massive 3-D storage lot. The interesting element in this process is the temporary existence as a *moving 1-D lot*: the stream-of-matter emerging from/through a 1×0.4-m opening at the bottom of the hopper, gradually emptying the entire hopper load of 10 ton soy beans. The complete ship's cargo will flow through this orifice. While the ship's cargo is rightfully characterised as an "impossible-to-sample" 3-D lot, the hopper outflow makes all the difference in the world. Imagine that a "cross-stream cutter" (illustrated by the principal sketch in Figure 8.3)

Figure 8.3. Often an "impossible-to-sample 3-D lot" will at some point be in a state of transportation—which then by definition constitutes a 1-D lot, a stream-of-matter that can be intercepted by incremental sampling. The top illustration shows a surprising parallel between a cross-stream sampler and a moving fish elevator (transportation from the cargo hold of a trawler). While representative sampling from the full trawler cargo hold is impossible, the fish elevator constitutes a perfect location for incremental sampling able to "cover" the entire cargo hold as it passes by. It matters not that the "particles" in this example are unusually large with respect to the unit increment volume—proper sampling also gets done here, but a relatively high number of increments will of course have to be used. The bottom illustration ("hopper-to-truck" discharge) is discussed in full in the text. Insert: sketch of the generic "cross stream cutter", or "cross stream sampler"); see also book front cover and Figure 8.4.

can easily be implemented at this location! This sampler will be able to cut correct, representative slices of this stream, i.e. correct increments. The only remaining question would actually be: how many increments are needed in order to characterise the entire ship's cargo?

8.3 Process sampling generalised

The scene is now set for a revelation. As soon as one has decided on honouring the full transverse coverage demand for every increment extracted, it is also clear that one can always cover the elongated dimension (the defining 1-D dimension) fully. This is simply a matter of covering the entire extension of the lot, whether by walking up the full length of the stationary lot, or if the dynamic lot, conveniently, streams past your sampling location. Where one is at liberty to choose, there is no doubt which situation would be preferred—it is indeed a very convenient situation, simply repeating the correctly cross-cutting increment sampling Q times. This type of sampling is, of course, easiest if *automated*, giving absolute sampling powers over all forms of flowing streams of matter, compare Figures 8.1 and 8.4.

8.4 Q

If the objective of the sampling is to produce a representative average analytical estimate pertaining to the whole lot, we

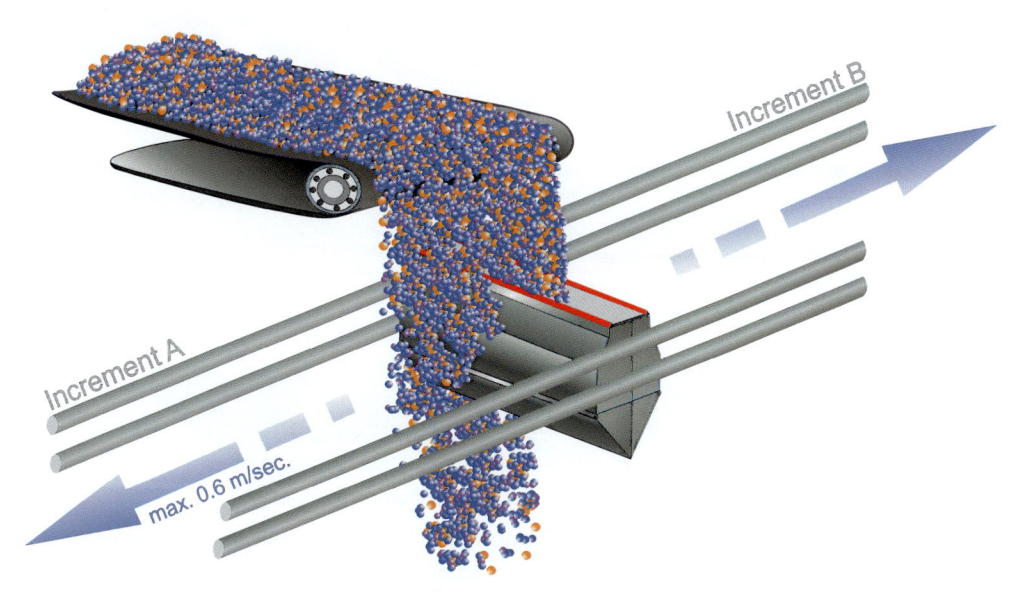

Figure 8.4. Principal sketch of archetypal cross-stream cutter for 1-D lot sampling. This sampler, in principle, works for all types of aggregate materials, and sometimes for slurries as well—but both application realms should always be subject to proper performance validation, see chapter 9. Illustration courtesy of Mr Martin Lischka, with permission.

observe that after Q increments extracted in this fashion, we have indeed covered all three dimensions of the original lot. All that was needed was to intercept the lot while it was in a moving state (1-D), see Figure 8.5.

Upon reflection there are few lots (if any) that are created in a finished state as 3-D lots. It is rather the case that these "big, impossible-to-sample 3-D lots" are *created* by a process meticulously laying up (stacking) a series of units. For example, units being delivered at the terminal end of a conveyor belt, or units delivered through a pipeline or otherwise. Imagine how easy the job would be if one *always* were able to install a relevant variant

Representative incremental 1-D lot sampling

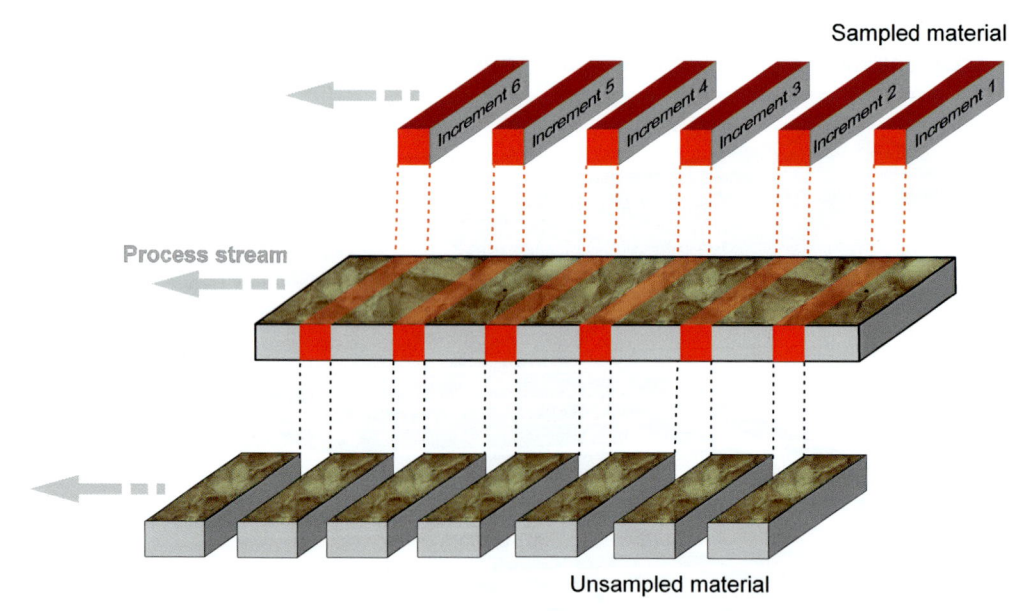

Figure 8.5. Fully representative incremental sampling of a 1-D lot. This illustration serves as a "role model" for correct transverse increment delineation and extraction which will have many different manifestations in real-world examples, but all should comply with the principle illustrated. In practice, all increments extracted from a *moving* 1-D lot will outline an *oblique* trace across the width of the stream, which has the same correctness as the principal orthogonal traces shown here. Illustration courtesy of Mr Martin Lischka, with permission.

Composite Sampling II: Lot Dimensionality Transformation

of a cross-stream sampler at the terminal end of a conveyor belt intercepting the falling stream of matter that is slowly building up the 3-D lot-to-be. This situation is rightly seen as the prototype procedure that can be turned into almost any composite sampling scheme desired—and always at the sampler's leisure, see Figure 8.7. So, all that is left is now is: how many increments to accumulate? The answer is simple—Q!

It goes without saying that the TOS owes the sampler an answer to the fair question: "HOW DO I ESTIMATE Q?", and an answer will be given, but spread across later chapters. All the professional sampler needs to do is never relinquish the composite

KeLDA study
Anticipating initiation to variographic characterisation in the next chapters, the KeLDA study is a showcase of practical process sampling issues in combination with a significant TAE of ~15–20%. References 1–3 apply variographic data analysis to this very complex data set forcefully illuminating many key TOS concepts.
See 👆 bit.ly/tos8-0, http://bit.ly/tos8-2, http://bit.ly/tos8-3

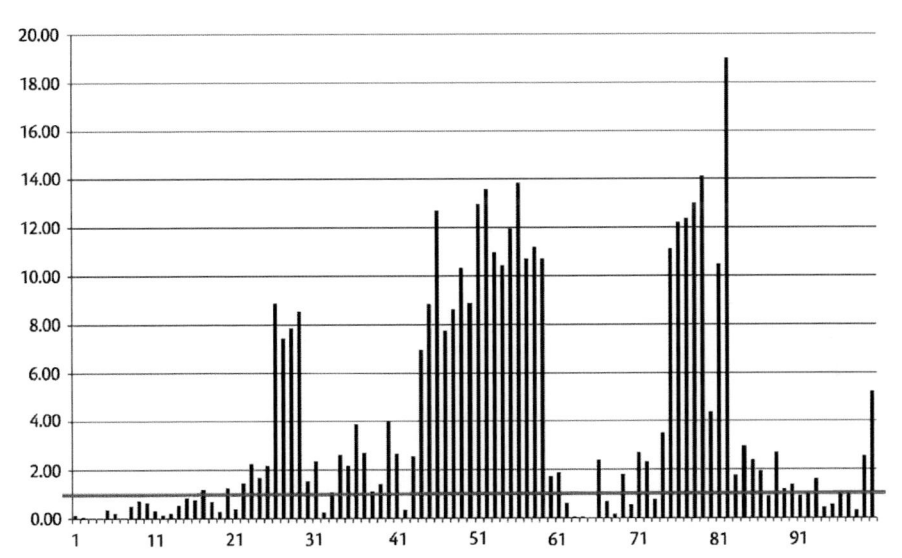

Figure 8.6. Representative extraction of 100 increments of a 1-D process stream revealing a pressing need for an effective composite sampling scheme if one is to be able to state with any reliability an average concentration of the full lot presented. It turns out that the pertinent Q for this difficult task is $Q = 42$. This example derives from the famous KeLDA study which is reported in full in References 1–3. The grey line represent a regulatory authority threshold (concentration = 0.9) below which resides "effectively zero concentrations". N.B. This particular lot has been suggested as a candidate for "one of the world's most heterogeneous lots". Consider the 3-D heterogeneity of the original lot from which this is an off-loading "linearisation"—both in the 3-D as well as in the 1-D manifestation, this lot indeed displays heterogeneity writ large.

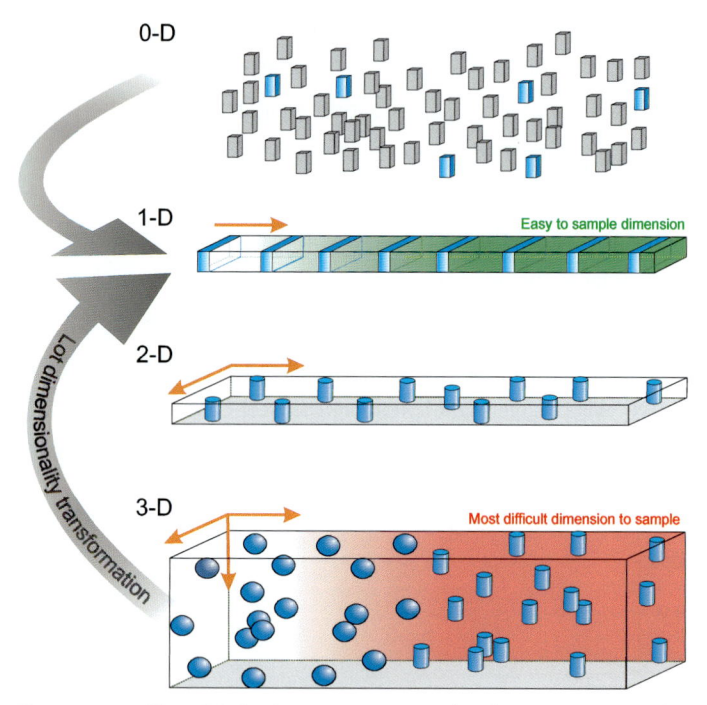

Figure 8.7. The TOS' Governing Principle: "Lot Dimensionality Transformation (LDT)".

sampling imperative. An example from a complicated system is given in Figure 8.6.

This chapter provides a first illustration of the framework in which composite sampling can be implemented without undue problems; it is all about the practicality of installing an appropriate cross-stream sampler (and *perhaps* also the costs involved). Although the latter must never be allowed to take over completely—the crucial primary sampling representativity is the only legitimate consideration! As always, if a primary sample is not representative—what is the meaning?

8.5 Lot dimensionality transformation (LDT)

The above serves to illustrate why the composite sampling imperative, together with a natural wish to conduct sampling in as easy a manner as possible, has led to a universal desire to hunt for all, or as many as possible, situations in which lots are in a similar state of 1-D transportation; or can be forced into such a situation without "too much" effort. There is great power in realising how immensely easier sampling can be if achieved from a 1-D lot. Later we shall also see why this is actually the situation in which the most effective sampling can be achieved, especially with respect to the critical possibilities of eliminating the adverse sampling bias.

This situation has been codified as one of the six Governing Principles of TOS: "Lot Dimensionality Transformation (LDT)", which plays a fundamental role in the primary sampling arena (Figure 8.7).

8.6 References

1. K.H. Esbensen, C. Paoletti and P. Minkkinen, "Representative sampling of large kernel lots – I. Theory of Sampling and variographic analysis", *Trends Anal. Chem.* **32**, 154–165 (2012). https://doi.org/10.1016/j.trac.2011.09.008, bit.ly/tos8-0
2. P. Minkkinen, K.H. Esbensen and C. Paoletti, "Representative sampling of large kernel lots – II. Application to soybean sampling for GMO control", *Trends Anal. Chem.* **32**, 166–178 (2012). https://doi.org/10.1016/j.trac.2011.12.001, bit.ly/tos8-2
3. K.H. Esbensen, C. Paoletti and P. Minkkinen, "Representative sampling of large kernel lots – III. General Considerations on sampling heterogeneous foods", *Trends Anal. Chem.* **32**, 179–184 (2012). https://doi.org/10.1016/j.trac.2011.12.002, bit.ly/tos8-3

9 Sampling quality assessment: the replication experiment

This chapter gives an overview of an issue that has not received proper attention, "replication". This issue turns out to be complex and there has been a lot of confusion in the literature and among professional colleagues. Three answers to what is often stated in response to the fundamental question: "what is *replicated* here exactly?" are i) replicate samples, ii) replicate measurements or iii) replicate analysis (replicate analytical results). Upon reflection it is clear that these three answers are not identical. The often only *implied* understanding for all three cases is that a beneficial *averaging* is carried out with the connotation that important reduction of uncertainty can be gained by "replication". By repeating the specific process behind replicated samples, measurements or results, some measure of variability is obtained; but a measure of what? There are many vague prerequisites and imprecise assumptions involved, which need careful analysis. For starters, i) addresses the **pre**-laboratory realm, while ii) and iii) play out their role **in** the analytical laboratory—but even here: is replicate analysis the same as replicate measurements?

AA
DOE: Design of Experiments
TAE: Total Analytical Error
TSE: Total Sampling Error
GEE: Global Estimation Error
IDE: Increment Delineation Error
IEE: Increment Extraction Error
MU: Measurement Uncertainty
PAT: Process Analytical Technology
RE: Replication Experiment
RSV: Relative Sampling Variability
STD: Standard Deviation

9.1 Background
The discipline of Design of Experiments (DOE) employs a strict conceptual understanding and terminology because of the controlled conditions that can be attained. In the situation of chemical synthesis influenced by several experimental factors such as

temperature, pressure, concentration of co-factors for example, it is easy to understand what a replicate experiment means: one is to repeat the experimental run(s) under *identical* conditions for all controllable factors, while taking care to randomise for all other known but uncontrolled factors. In this case the variance of the repeated outcome, be it small or large, will furnish a measure of the "total experimental uncertainty". In routine operations in the analytical laboratory, variability also reflects effects from other uncertainty contributions, for example stemming from small-scale sampling of reactants involved, which may not necessarily represent completely "homogeneous stocks". Added uncertainty contributions may also occur from resetting the experimental setups, e.g. to what precision can one "reset" temperature, pressure, concentration levels of co-factor chemical species after having turned off and cleaned the experimental equipment before resuming experimentation? Still, such uncertainty contributions are usually considered acceptable parts of the total analytical error (TAE). Often, all of the above turn out to be of small, or vanishing, effect because of the regular conditions surrounding a controlled DOE situation.

However, going back one step, one might find it equally relevant to repeat the experiment by another technician, researcher and/or in another laboratory—enter the realm of the well-known analytical concept of **reproducibility**. There may be more, smaller or larger, effects in this widened context, and careful empirical total effect estimations must always be carried out in order to arrive at a valid estimate of the augmented, effective TAE.

9.1.1 Behold the whole lot-to-analysis pathway

Below we address still more external issues, not always on the traditional agenda for "replication", in fact quite often left out, or forgotten. But these are nevertheless often of crucial importance.

There are in fact many scenarios that differ from a nicely bracketed DOE situation. Indeed, most data sets do not originate exclusively from within the complacent four walls of an analytical laboratory. What will be described below constitutes the opposite end of a full spectrum of possibilities in which the researcher/data analyst must also acknowledge *significant* sampling, sub-sampling, sample preparation and other errors in addition to the effective TAE. The total sampling error (TSE) will include all sampling and mass-reduction error effects incurred *before* analysis. It is self-evident that these errors must also be included in *realistic* analytical error assessments; TAE alone will not give a relevant, valid estimate of the total effects influencing the analytical results. We are forced to furnish a valid estimate of the *total* sampling–sub-sampling–sample preparation–analysis uncertainty estimate, the Global Estimation Error (GEE = TSE + TAE).

The description below intends to deal comprehensively with the many different manifestations surrounding the replication issue, such that most realistic scenarios are covered. At the heart-of-the-matter is a key question: what is meant by "replicate samples"? This issue is actually more complex than may appear at first sight, and will receive careful attention w.r.t. definitions and terminology. It will also transpire that this issue is intimately related to *validation* in data analysis, chemometrics and statistics, if the reader has reason to invoke such disciplines when addressing analytical results.

9.2 Clarification

Upon reflection it will be appreciated that the process of "replication" can concern the following alternatives in the full lot-to-aliquot pathway:

1) Replication of the primary sampling process

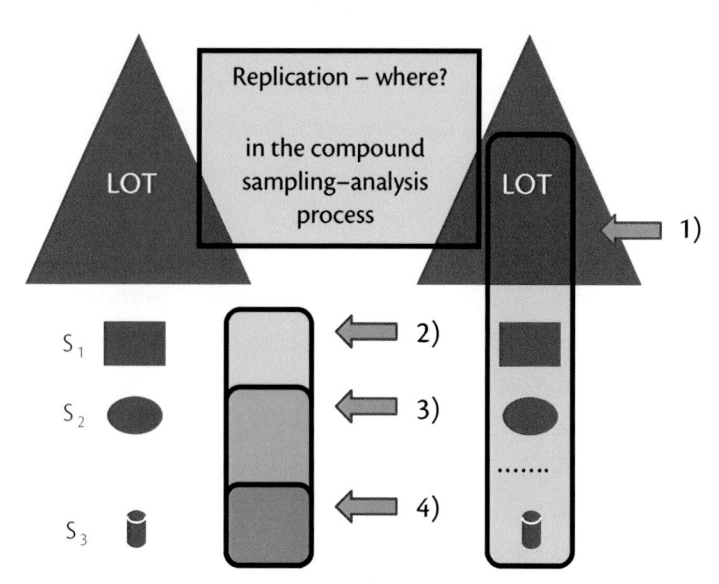

Figure 9.1. Replication can be performed at many stages in the full lot-to-aliquot pathway, but which is the most *realistic* situation pertaining to the general operations not only in the analytical laboratory? It turns out that all replication **must** meaningfully start "from the top", i.e. from option 1) representing sampling from the full lot.

2) Replication starting with the secondary sampling stage (i.e. first mass reduction)
3) Replication starting with the tertiary sampling process (i.e. laboratory mass reduction)
4) Replication starting with aliquot preparation, where material spillage may occur, for example: IDE/IEE
5) Replication starting with aliquot instrument presentation, e.g. surface conditioning
6) Replication of the analysis (measurement operation) only (TAE)

The last option is the situation corresponding to "replicate measurements" in the most restricted case, which is supposed to allow to estimate TAE. However, does this mean that the analytical aliquot (the vial) stays in the analytical instrument all the time while the analyst simply "presses the button" say 10 times? Possibly; in which case this furthers a strict estimate of TAE *only*. However, it seems equally *relevant* to extract the vial and insert it in the instrument repeatedly, allowing a possible temperature variation to influence replicated measurements—because this is a more *realistic* repetition of the general work and measurement process in the laboratory than simply leaving the test portion in the instrument. This is a first foray into what is known as "Taguchi thinking",[1] which focuses on potentially influencing factors which are not embedded in the experimental design explicitly. Clearly this kind of external thinking is relevant in many situations and should therefore be included in the replication approach. One important dictum of Taguchi is: do not necessarily look only for optimal results (which *may* have large variability), but to results where the response variability is low over a large span of the experimental domain (even if less optimal). This is a clever way of gaining more information about the measurement process involved in a more realistic context, be this a production or manufacturing process, or the analytical process itself. Certain scepticisms have been voiced regarding the merits of this approach, but we will let the reader decide on this matter by reading on with an open mind.

Starting with this type of *perturbation* of the analytical process elements, for another analyst it may appear equally reasonable to include some, or all, of the "sample preparation" procedures in the replication procedure as well, because these part-operations cannot necessarily be performed in a completely identical fashion either. This effect should then also be repeated, say 10 times (stages 4 and/or 5 above) in order also to *include* a measure of this particular variance contribution.

But having broadened the horizon thus far, it is a completely logical step to follow up with still further realistic perturbations of the measurement process, which broadly means also including the tertiary, secondary and in the full measure of things, even also primary sampling errors in the replication concept. Why? Because these are *de facto* uncertainty contributions that will be in play for any-and-all analytical aliquots ever subjected to measurement! Following the full impact of the TOS and its detailed treatment of the phenomenon of *heterogeneity*, it is clear that the only complete "sampling-and-analysis" scenario that is guaranteed to include **all** uncertainty contributions **must** start with replication of the primary sampling ("replication from the top"). Any less comprehensive replication scenario is bound to be incomplete.

Repeating the primary sampling, again at least 10 times (preferentially more if/when needed), means that each of the individually sampled primary samples is subjected to an identical protocol that governs **all** the ensuing sub-sampling (mass-reduction), sample handling and preparation stages, and procedures in the laboratory. Logically this is the only approach that incorporates the complete ensemble of errors incurred from-lot-to-analysis, of whatever nature (sampling, sub-sampling, handling, preparation, presentation). The point is that for each replicated primary sample, all potential errors will be manifested *differently* ten individual times giving rise to an accumulated variance which is the most realistic estimate of the **total** measurement uncertainty (MU).[2] In particular, this estimate is bound to include the full sampling error effects (TSE), which will often dominate.

In clear contrast, starting at **any** other of the levels in the list above, stages 2–6 will guarantee an incomplete, inferior TSE + TAE estimation, which is structurally destined to be too low, i.e. unrealistic.

Should one nevertheless feel compelled to "shortcut" the full replication procedure by not starting "from the top", one is **mandated** to describe the rationale behind this choice, and to provide a **full report** of what was in fact done, lest the user of the analytical results has no way of knowing what was implicated in the umbrella term "replication". Users and decision makers, acting on the analytical data which are supposed to be reliable, do not like to be kept in the dark.

Undocumented or unexplained application of the term "replicate experiments" (or "repeated experiments") has been the source of a significant amount of unnecessary confusion. Many times s^2(TAE) has simply been *misconstrued* to imply s^2(TSE + TAE), a grave error, for which *someone* or *somebody* (or some ill-considered, incomplete protocol) is responsible! But here we are not interested in pointing fingers at any entity (private or legal); it suffices to stop continuing with such misguided practice.

The above scenario illustrates the unfortunate responsibility compartmentalisation, which is rather often found in scientific, industrial, publishing or regulatory contexts; here are a few examples:

"The analyst is not supposed to deal with matters *outside* the laboratory (e.g. sampling)"

"This department is *only* charged with the task of reducing the primary sample to manageable proportions, as per codified laboratory instructions"

"Sampling is automated, and carried out by process analytical technology (PAT) sensors; there is no sampling issue involved here"

"I am not responsible for sampling, I only analyse/model the *data*"

... and many similar *excuses* for not seeing the complete measurement uncertainty context. All too often the problem belongs to "somebody else", with the unavoidable result that the problem does not receive further attention. This stand ("not my responsibility") is, therefore, always potentially in danger of being perpetuated—and whenever so, "replicate analysis" will still take its point of departure at stage 3 (maybe stage 2), but never from stage 1, the primary sampling stage. This is not an acceptable situation. There are many occasions in which authors, reviewers and editors have missed cracking down with the necessary firmness on such demonstrable ambiguities regarding "replication", with the certain result that the reader is not able to understand what was intended, nor what was indeed carried out, because of incomplete descriptions in the "Method" sections of scientific publications and technical reports. The issue is far from trivial, indeed grave errors are continuously being committed. But rather than address the obvious question (*who* is responsible), the way forward is here to be constructive. The focus shall rather be on ways and means to put an effective end to the confusion surrounding the replication.

9.3 Quantifying total empirical variability—the replication experiment

How a realistic estimate of the *total* TSE + TAE, a replication experiment (RE) must always start "from the top" was outlined above. This is where replication starts, be this primary sampling in nature, in the field, sampling in the industrial plant or sampling of any target designated as the primary lot (examples follow below).

Figure 9.2 shows the scenario where an avid sampler is facing a large lot with the objective of establishing a realistic estimate of the average lot concentration for one analyte (or more). It is abundantly clear that a single grab sample stands no chance of ever being able to do this job because of the intrinsic distributional heterogeneity of the lot. It does not matter whether the lot is small, intermediate or large, the point being that this intrinsic heterogeneity is *unknown* at the moment of sampling. The sampler, therefore, has no other option than to act **as if** it is significantly large. There is no problem assuming this rational stance,

Sampling Quality Assessment: the Replication Experiment 101

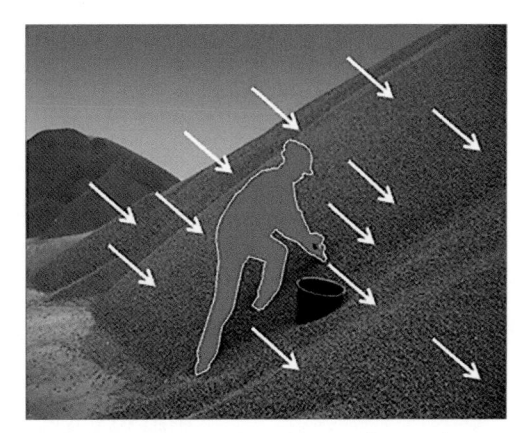

Figure 9.2. A primary sampler approaching a significantly heterogeneous lot with a grab sampling RE approach, but deployed with two very different *footprints*. The sampler on the left approaches the RE on an irrationally narrow footprint in relation to the full geometrical scale of the lot. The sampler on the right attempts to take account of the (hidden) lot heterogeneity by employing a wider footprint as a basis for the RE. These alternative scenarios will result in *different* relative sampling variability estimates because of the different lot heterogeneities covered. (N.B. **neither** of these primary sampling procedures succeeds to sample the interior of the lot, so both do **not** honour the fundamental sampling principle (FSP).

the TOS furnishes all necessary governing principles and practical procedures and equipment assessment possibilities so as always to be able to deal with significant lot heterogeneity, e.g. Esbensen & Julius.[3]

By deploying a RE (Figure 9.2, right), the sampler now has access to a first estimate of the effective variability of the sampling procedure, but with the TOS it is also clear that there is still a grave breach of the FSP. A proper RE must honour the full FSP demands of covering the entire lot volume.

9.4 Relative sampling variability

It has been found useful to employ a general measure of the sampling variability, as expressed by a RE—enter *RSV*: the Relative Sampling Variability.

The variability between any number of replications can be quantified by extracting and analysing the set of analytical results from a number of replicate primary samples. These specifically should aim to *cover* the entire spatial geometry of the lot *as well as possible*, i.e. spanning the geometrical volume of the primary lot in an optimal fashion (given the circumstances), and calculating the resulting empirical variability based on the resulting analytical results a_S. Often a relatively small number of primary samples may suffice for a first survey, though never less than 10. It is essential that the sampling operations are fully realistic replications of the standard routines, i.e. they shall **not** be extracted at the same general location, Figure 9.2 (left), which would only result in a *local* characterisation not at all able to represent the effects of the full lot heterogeneity. What is meant here is that the successive primary sampling events shall take place at other, *equally likely* locations, where primary sampling could take place. The RE shall be carried out according to a strict protocol that specifies precisely how the following sub-sampling, mass reduction and analysis steps are to be carried out. It is essential that both primary sampling as well as all sub-sampling and mass-reduction stages and sample preparation is replicated in a completely identical fashion in order not to introduce unnecessary extra, irrelevant variability in the assessment.

Note that when these stipulations are followed it is possible to conduct a RE for any sampling procedure, for example a grab sampling, or for a composite sampling procedure or, for that matter, for **any** other sampling procedure.

It has been found convenient to employ a standard statistic to the results from a RE. The relative coefficient of variation, CV_{rel} is an informative measure of the relative magnitude of the standard deviation (STD) in relation to the average (X_{avr}) of the replicated analytical results, expressed as a %:

Sampling Quality Assessment: the Replication Experiment

$$CV_{rel} = \left[\frac{STD}{X_{avr}}\right] \times 100 = RSV \qquad (9.1)$$

RSV is called the Relative Sampling Variability (or relative sampling standard deviation). *RSV* encompasses all sampling and analytical errors combined, as manifested by a minimum 10 times replication of the sampling process being assessed. *RSV*, therefore, measures the total empirical sampling variance influenced by the specific heterogeneity of the lot material, *as expressed by the current sampling procedure*. This is a crucial understanding. There can be no more relevant summary statistic of the effect of repeating the full lot-to-aliquot sampling procedures (10 or more times) than a RE-based *RSV*.

In the last decade there has been a major discussion in the international sampling community as to the usefulness of a singular, canonical *RSV* threshold; opinions have been diverse. In the last few years a consensus has emerged, however, that *indicates* a general acceptance threshold of 20%. *RSVs* higher than 20% signify a too-high sampling variability, with the consequence that the sampling procedure tested **must** be improved so as better to counteract the inherent heterogeneity effects in the lot material. Should one elect to accept a *RSV* higher than 20%, for example, this will have to be justified and made public to ensure full transparency for all stakeholders.

The usefulness of the *RSV* measure cannot be underestimated. For *whatever* lot material, sampled by *whatever* procedure, the specific lot/procedure *combination* can be very quickly assessed and characterised by a specific *RSV*. There are no untoward practicalities involved which might militate against performing a RE assessment; indeed, anybody can perform RE assessment on any sampling procedure, or for any sampling equipment. It should thus never be possible to argue for, or against, a specific sampling procedure without a transparent quantitative assessment in the

Beware of false universalities
DS 3077 offers this critical caution: "It is obvious that a single universal RSV threshold that is supposed to apply to all types of materials and lots interacting with very different sampling processes is too much to wish for. This violates the very nature of heterogeneity... There is clearly a limit to the validity of a completely general threshold. ... It is recommended that end-users educate themselves so as to be able to decide on a relevant level." Reference 4, p. 24.

form of a relevant *RSV*. RE numbers speak for themselves. The "difficult" issue of sampling is put on a fully understandable and very simple assessment basis—the RE.

Based on extensive practical experience over 50 years from many applied sectors and fields within science, technology and industry, there are very many cases on record in which the 20% threshold is exceeded (not infrequently by significant deviations); but there are also an important number of cases in which the existing procedure is vindicated. A few illustrative examples are given below. But first: what information resides in a simple *RSV* level?

Figure 9.3 illustrates how *STD* can be expressed as a *fraction* of the general concentration level quantified by X_{avr}. In this illustration, cases are shown where the empirical *STD* forms, e.g. 10%, 25%, 50%, 65%, 85%... The issue clearly is, at what %-level is one no longer comfortable with the quantification resolution, e.g. for

A universal Threshold (RSV %)?

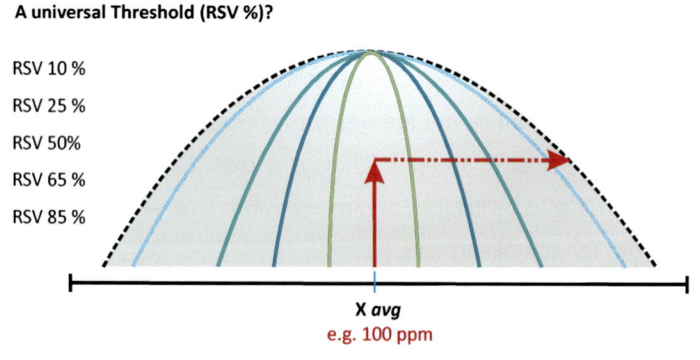

RSV 10%
RSV 25%
RSV 50%
RSV 65%
RSV 85%

X avg
e.g. 100 ppm

Figure 9.3. Schematic illustration of replication experiment thresholds RSV, 10%, 25%, 50%, 65% and 85%. Very large relative standard deviations (higher than approximately 85%), when interpreted as representing a standard normal distribution, apparently *may* give rise to negative concentration values. This has no physical meaning, however, and need not cause any worry; these are but model fitting artefacts, of no practical consequence. The essential information for the sampler is manifest already when RSV transgresses >20%, i.e. when the sampling procedure is operationally too variable and must be improved upon (the TOS).

$RSV = 50\%$ the signal-to-noise ratio is 1:1, which is absolutely an unacceptable situation under any circumstances.

The canonical *RSV* threshold level, 20%, serves as a general indication **only** in the case where *nothing* is known *a priori* as to the heterogeneity of the material involved. Materials and material classes certainly exist that may merit a higher, or a lower, threshold, for which the proposed general *RSV* value can no longer serve. For such cases, a material-dependent quantification will have to be developed, dependent upon the sampler's own competence and diligence. The mandate in the sampling standard DS 3077[4] is clear: **all** analytical results must henceforward be *accompanied* by an appropriate *RSV*, voluntarily described and reported in full.

While it is acceptable to level criticism against a suggested general threshold (20%), this also entails the obligation to perform empirical due diligence in the form of a RE. Recent industrial, scientific and technological history is full of examples of major surprises brought about by such simple replication experiments and their attendant *RSV*. When- or wherever the RSV threshold of 20% is transgressed, it is either the intrinsic material heterogeneity which is underestimated or the sampling procedure turned out to be much less fit-for-purpose than assumed. There are no other possible conclusions in such a case.

The purpose of a RE is often to assess the validity of an already existing sampling and analysis procedure. In practice, the RE can only perform and test a current sampling procedure as it *interacts* with a specific lot material. *Should* a *RSV* for this exploratory survey exceed the canonical, or case-specific, threshold, the need for complete compliance with the TOS has been documented and is, therefore, mandated, no exceptions allowed. There may be good reasons to start validation by testing an existing sampling procedure—there is always the *possibility* it may turn out to fall

Figure 9.4. Examples of replication experiments (RE) that are easily set up. Right: a dynamic process sampling situation. Left: sampling from a stopped belt, i.e. a temporary stationary lot. Both sampling scenarios can be assigned an objective *RSV* quality index. In order that no misunderstanding may occur, it is usually only necessary to perform a proper, calibrating RE once, as part of surveying and characterising the intrinsic heterogeneity of a *specific* lot material. Should the sampling target experience significant variations, however (e.g. over time, change of raw material and/or supplier), it will be prudent to repeat a basic RE; this does not require much, only 10 primary samples.

below the pertinent threshold, and thus be acceptable as is. But in all other cases, TOS-modifications **must** be implemented, no exceptions.

Therefore, one can view *RSV* as a flexible and relevant sampling procedure *quality index*, scaled with the inherent heterogeneity encountered. *RSV* is particularly useful for initial characterisation of sampling from *stationary lots*, while it is much more customary to use a dynamic, process sampling augmented approach, called *variographics* when sampling from dynamic lots. *RSV* and *variographics* are closely related approaches for quantifying heterogeneity; the latter approach is much more powerful, due to the fact of its more elaborate experimental design which allows full decomposition of GEE, see, for example, References

Sampling Quality Assessment: the Replication Experiment

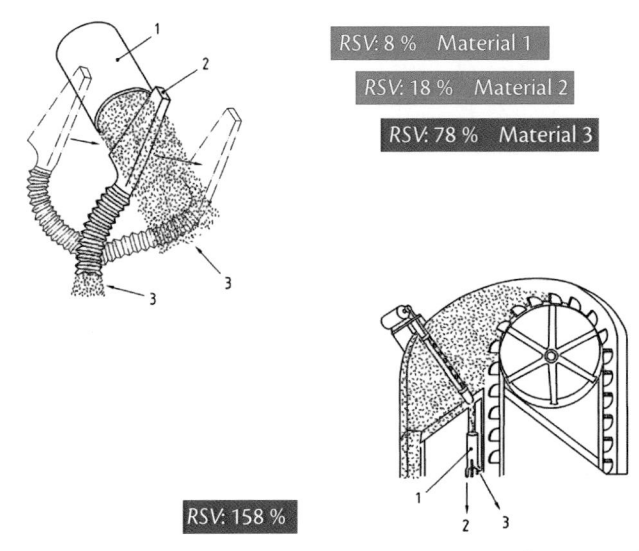

RSV: 8 %   Material 1

RSV: 18 %   Material 2

RSV: 78 %   Material 3

RSV: 158 %

Figure 9.5. Upper left: primary process sampler assessed for three different materials, one of which does not pass the test of the dedicated RE (*RSV* = 78 %). Lower right: a complex primary sampler being subjected to a RE with the distinctly worrisome result of *RSV* = 158 %. N.B. illustrative examples only, no specific sampler is endorsed, nor renounced. Samplers are sketched only in order to illustrate how RE may be used for quantitative assessment.

High-level example
For the specially interested: "The advantages and pitfalls of conventional heterogeneity tests and a suggested alternative",[9] (👆 bit.ly/tos9-9) will propel the reader to the frontline of contemporary TOS deliberations, a.o. showcasing several, complex embedded RE issues. This advanced example also introduces the Poisson process viewpoint, which is necessary to understand many of the more subtle issues in sampling.

5–8. Variographic heterogeneity characterisation of dynamic lots is the subject of later chapters.

All examples described above pertain to issues related to sampling and other error contributions *before* analysis. It is noteworthy that some analytical procedures can have significantly large TAE, e.g. of the order of 10–20 % or more, which is then already factored into the empirical *RSV* level. The principal issues from the few examples given here can be generalised to many other material and lot types. The GEE = TSE + TAE issues are identical for all lot systems.

The examples in Figure 9.5 illustrate how one specific sampling equipment can be assessed with respect to different

What a wonderful world this could be...
Imagine that all the world's analytical results from this date on were compelled also to present the relevant RE measure, e.g. 57.3 ppm ± *RSV*(%). This would allow all users and decision makers to assess the validity of the full TSE + TAE pathway—what could be more relevant? This is exactly what is recommended in DS 3077![4]

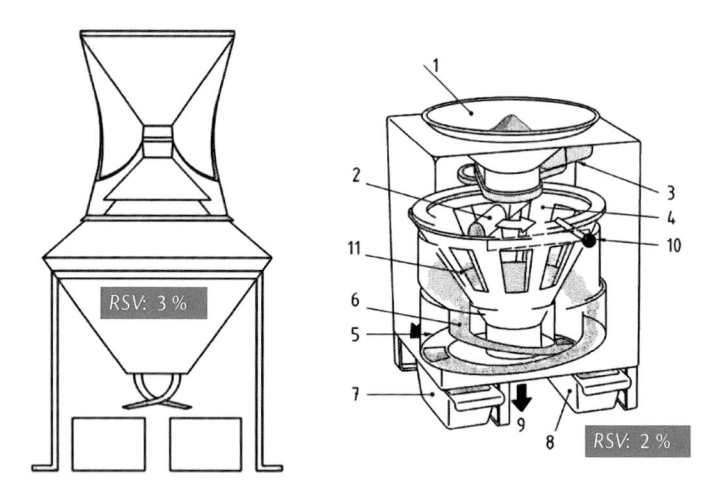

Figure 9.6. Two laboratory equipment (splitters) subjected to RE assessment, showing highly satisfactory quantitative results. N.B. illustrative examples only, no specific sampler is endorsed, nor renounced. Samplers are sketched only in order to illustrate how RE may be used for detailed quantitative assessment in the sub-sampling, -preparation laboratory realm.

RE to the extreme
A special development study, "Single kernel NIR analysis of bulk wheat heterogeneity—a Theory of Sampling reference study" by E. Tønning et al., contains a detailed breakdown of the entire sampling + analysis pathway illuminating how to achieve the smallest TSE in combination with a significant TAE. This study systematically applies a hierarchical RE which results in new insight into the pre-NIR process, previously unknown: http://bit.ly/tos9-b

---

materials (with specific heterogeneities), which may result in both pass and fail.

RE is a general facility that can be deployed at all stages in the lot-to-aliquot pathway, i.e. also from stages later than the primary sampling stage. If the objective were to assess and compare the two splitters in Figure 9.6 specifically, the RE may well be initiated at this sub-sampling stage directly.

The replication experiment (RE) is a powerful and highly versatile sampling quality assessment facility that can be deployed with great flexibility. It is necessary to be fully transparent as to what is meant by "replication" in all particular situations, for example at what stage in the lot-to-analysis pathway is replication to commence.

The reader is recommended to peruse and contemplate Reference 2 carefully. For the interested reader this reference furthers more in-depth viewpoints and is essential background for the entire book.

9.5 Notes and references

1. Wikipedia, *Taguchi Approach*. http://en.wikipedia.org/wiki/Taguchi_methods, 👆 bit.ly/tos9-1
2. Issues related to the concept of Measurement Uncertainty (MU), which too often in practice only covers the parts of the analysis process that can be brought under direct laboratory control (while in its full definition purports to cover the entire sampling-handling-analysis pathway), are treated in: K.H. Esbensen and C. Wagner, "Theory of Sampling (TOS) versus Measurement Uncertainty (MU) – a call for integration", *Trends Anal. Chem.* **57**, 93–106 (2014). https://doi.org/10.1016/j.trac.2014.02.007, 👆 bit.ly/tos1-6
3. K.H. Esbensen and L.P. Julius, "Representative sampling, data quality, validation – a necessary trinity in chemometrics", in *Comprehensive Chemometrics*, Ed by S. Brown, R. Tauler and R. Walczak. Elsevier, Oxford, Vol. 4, pp. 1–20 (2009).
4. DS 3077. *Representative Sampling—Horizontal Standard*. Danish Standards (2013). http://www.ds.dk, 👆 bit.ly/tos1-2
5. K.H. Esbensen, A.D. Román-Ospino, A. Sanchez and R.J. Romañach, "Adequacy and verifiability of pharmaceutical mixtures and dose units by variographic analysis (Theory of Sampling) – A call for a regulatory paradigm shift", *Int. J. Pharm.* **499**, 156–174 (2016). https://doi.org/10.1016/j.ijpharm.2015.12.038, 👆
6. K.H. Esbensen, C. Paoletti and P. Minkkinen, "Representative sampling of large kernel lots – I. Theory of Sampling and variographic analysis", *Trends Anal. Chem.* **32**, 154–164 (2012). https://doi.org/10.1016/j.trac.2011.09.008, 👆 bit.ly/tos8-0

7. K.H. Esbensen, C. Paoletti and P. Minkkinen, "Representative sampling of large kernel lots – III. General Considerations on sampling heterogeneous foods", *Trends Anal. Chem.* **32,** 178–184 (2012). https://doi.org/10.1016/j.trac.2011.12.002, 👆 bit.ly/tos8-3
8. P. Minkkinen, K.H. Esbensen and C. Paoletti, "Representative sampling of large kernel lots – II. Application to soybean sampling for GMO control", *Trends Anal. Chem.* **32,** 165–177 (2012). https://doi.org/10.1016/j.trac.2011.12.001, 👆 bit.ly/tos8-2
9. A high-level example for the specially interested: F. Pitard, "The advantages and pitfalls of conventional heterogeneity tests and a suggested alternative", *TOS Forum* **5,** 13–18 (2015). https://doi.org/10.1255/tosf.47, 👆 bit.ly/tos9-9

10 Sampling quality criteria

Formulating proper Sampling Quality Criteria (SQC) is the initial step in a scientific approach to representative sampling. This activity can be characterised as "a framework for planning and managing sampling and analytical operations consistent with the overall project objectives". It includes establishment of concise sampling objectives, precise outlining of a decision unit (DU) and deciding on the level of confidence wanted regarding the kind(s) of decisions to be made based on the analytical results. Once defined, these criteria serve as input to the Theory of Sampling for developing a representative sampling protocol.

SQC: Sampling Quality Criteria
DU: Decision Unit
FSP: Fundamental Sampling Principle
GEE: Global Estimation Error
TSE: Total Sampling Error
PAT: Process Analytical Technology

10.1 Sampling quality criteria

The first component in SQC is defining the *analyte(s)* to be involved, the concentration ranges of interest of the analyte(s) and how inference(s) will be made from the analytical data to the decision unit.

The second component of the SQC concerns definition and establishment of the physical *decision unit* (DU), also known as the "lot" in the TOS. The decision unit establishes the spatial and temporal boundary conditions of the sample collection process.

The third SQC component, the confidence level, establishes the desired probability with which a correct inference (decision) *can* be made. The confidence level should typically correlate to the potential consequences of an incorrect decision, e.g. as concerns health, economic, environmental or other societal consequences. The magnitude of the total combined error effect in the sampling, sample processing and analytical protocols (TSE + TAE) constitute the unavoidable basic risk level involved, and determines the likelihood of an incorrect decision. Thus, controlling errors to a greater extent increases the probability of a correct decision. The required confidence level also directly affects the sampling effort and quality control measures.

Establishing proper SQC is not difficult, it is nothing but very carefully thinking through the *why*, *what* and *how* regarding the use of the final analytical results; surprisingly, often this prerequisite does not get the full attention it deserves.

10.2 First SQC component—definition of analyte(s)

The first SQC component addresses definition of the analyte(s), the expected concentration level and range of interest, as well as how the analytical data will be used and interpreted: what information is expected? In other words, how *inference* is made with respect to the decision unit. This prerequisite sets the scope and limits the selection possibilities for sampling tools, sampling containers, sample handling procedures, sample preservation, laboratory preparation equipment, sample mass reduction procedures etc. For all sampling operations, it is critical to secure analyte integrity from the primary sample all the way through to the final analysis.[1]

10.3 Second SQC component—delineating the decision unit (DU)

The decision unit (*aka* the lot) defines the target (material, form, size, conditions) from which the primary sample is to be collected and importantly sets the *scale of decision-making*. This scale can be based on volume, mass, package size or any other definable criteria relevant for the project. As defined in a previous chapter, a pre-requirement for the decision unit is complete physical accessibility to the lot, aka the Fundamental Sampling Principle (FSP). If a certain part of the decision unit is not accessible structurally, this section either **must** be made accessible for the purpose of sampling, or the (limited) target is disqualified from being a proper decision unit from which defensible and reliable decisions can be made. Sampling of a decision unit under limiting constraints, be these practical, logistical, economic etc., will unavoidably forfeit any chance of extracting representative samples; only worthless specimens will ensue.

10.4 Third SQC component—inference and confidence

Three basic types of inference can be applied from the final analytical data, i.e. from the concentration of the analytes of interest: *judgement*, *direct inference* and *statistical analysis*.

Judgement is never an acceptable methodology, irrespective of whatever personal experience is involved. Judgemental inference is not discussed further here.

Direct inference is the simplest proper inference approach in which the analytical result from a single primary (composite) sample is used as a reliable, and defensible, estimate of the average concentration of the analyte in the full decision unit/lot. This approach requires no statistical analysis as long as the principles of the TOS are obeyed. This is the tacitly assumed sampling situation in many "straightforward" situations.

In comparison to direct inference, *statistical analysis* involves multiple primary samples. In such case, the upper confidence limit of the mean is compared to a specification limit, or a statistical comparison of one decision unit to another decision unit (e.g. reference decision unit) is carried out. In "acceptance sampling", e.g. in releasing a batch of drug dosages/tablets or in *similar* situations, it is the lower limit of the confidence interval that sets the operative threshold.

There is no rule for setting the confidence level. It is a function of the consequences of an undesired incorrect decision. Typically, the larger the consequences of an incorrect decision, the higher the desired confidence needs to be. It is **not** a good policy always to set the confidence at the same level (e.g., 95 %) if it is not known *why* this level is actually used. Never mind that this "usually" is the level encountered; there are enough "template statistics" governing complex problems, projects etc. It may just as well be that *careful* consideration (proper SQC deliberations) reveal that some other level of confidence is more appropriate.

In order to calculate a confidence level for statistical analysis, an estimate of the global estimation error (*GEE*) is required, as defined in previous sampling chapters:

$$GEE = Total\ Sampling\ Error + Total\ Analytical\ Error$$

For either case of direct inference or statistical analysis, the estimate of the *GEE* (all bias sampling errors must have been eliminated first, according to the TOS) can be determined from a properly conducted replication experiment, see previous chapter.

If the sampling bias errors have not been properly eliminated, the estimate of *GEE* will forever be of varying magnitude, with the consequence that the premise for a constant confidence level is broken.

Figure 10.1 graphically depicts the relationships between confidence, error and representativeness. Applying the TOS, including the defined sampling quality criteria, to any sampling protocol ensures representativeness, because the primary goal of the TOS is to minimise the total sampling error (*TSE*), which in turn increases the confidence of the inference made from the analytical data to the decision unit.

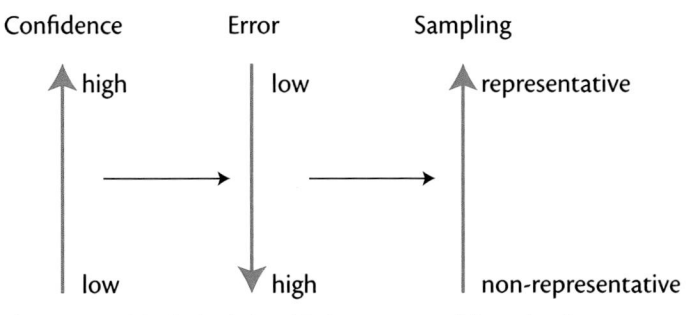

Figure 10.1. Principal relationship between confidence level, error magnitude and representativeness.

This chapter is only a brief introduction serving to raise the awareness of the need for proper SQC; the reader is referred to more in-depth treatments in the references,[2–7] with a natural starting point in References 1 and 6.

10.5   Perspectives

Science, technology, industry and society are critically dependent on the highest data quality (data relevance, data reliability, data representativity). Representative data cannot be acquired without a *sampling process* initiating the full "lot-to-analysis-to-decision" pathway. In this endeavour, the critical determinant is the potential *sampling bias*, an inconstant deviation between an analytical sampling result and the true average concentration of the lot, product or process, which must be *eliminated* from all sampling processes at all stages to ensure the expected, ideal data validity. The sampling bias is fundamentally different from the well-known analytical bias, which can be eliminated by a statistical bias correction. However, this approach is impossible for the sampling bias, which, in addition, is orders of magnitude larger in most cases. The sampling bias is treated in more detail below, see Chapter 23.

Sampling does not necessarily only refer to physical sample extraction, but also to on-line process and/or product measurements through the use of Process Analytical Technology (PAT) sensors. PAT is a framework originally developed by the US Food and Drug Administration to design, analyse and control pharmaceutical manufacturing processes through (continuous) measurement of critical process parameters, which is now being applied to many other manufacturing and processing industry sectors as well. If in such cases highly advanced measurement sensor technologies extract information from a wrongly defined decision unit, or from biased samples, the results of any designed process control technique will unavoidably lead to biased results. The main challenge is, therefore, to ensure that highly precise measurements indeed do reflect the true target value of the decision unit (requiring elimination of the sampling bias), which currently most of the times remains unaddressed. A fully comprehensive SQC is often a missing link in this context. A recent introductory treatment of the sampling bias in the context of PAT can be found in the textbook by Esbensen and Swarbrick,[8] and in Esbensen and Paasch-Mortensen.[9]

In all sampling situations (whether sampled and analysed physically, or measured by the use of PAT), *representativeness* (of samples and signals) is the prime objective without which the derived analytical results and the decisions based on these data may very easily be invalid. Representativeness implies both elimination of the sampling bias, as well as high reproducibility of the sampling process. Depending on the required use of data (used "as is" for monitoring or aggregate data for higher-level decision-making), there will always be a problem-dependent "decision unit" defining the target from which a sample is extracted or for which a signal is measured. But critical DUs do not always conveniently suggest themselves at "convenient" PAT sensor locations; problem-based due diligence is required!

10.6 Summary

Defining sample quality criteria must be the initial step in the development of any sampling protocol. SQC defines the analytes, the decision unit and addresses the required inference and its confidence level. The precise definition of the analyte(s) and the decision unit is currently one of the weaker elements, and not always sufficiently addressed, with the potential consequence of using improper sampling protocols, ultimately with the end result of invalid inferences. Reference 6 (freely available) is an indispensable follow-on read with which to get a more fully developed view of the critically important SQC issue.

An icon on the interface between statistics and the TOS, Mr Chuck Ramsey has been in the forefront of applied statistics for decades, insisting on incorporating the Decision Unit formalism, www.envirostat.org. He has also devised the ingenious Aloha sampler, see https://doi.org/10.1255/tosf.25, bit.ly/tos10-9

10.7 References

1. C. Ramsey and C. Wagner, "Sample quality criteria", *J. AOAC Int.* **98**, 265 (2015). https://doi.org/10.5740/jaoacint.14-247, bit.ly/tos10-1
2. K.H. Esbensen, *DS 3077 Representative Sampling—Horizontal Standard*. Danish Standards (2013). http://www.ds.dk, bit.ly/tos10-2
3. *GOODSamples: Guidance on Obtaining Defensible Samples*. Association of American Feed Control Officials, Champaign, IL (2015). http://www.aafco.org/Publications/GOODSamples, bit.ly/tos10-3
4. F.F. Pitard, *The Theory of Sampling and Sampling Practice*, 3rd Edn. CRC Press (2019). ISBN: 978-1-138476486
5. C.A. Ramsey and A.D. Hewitt, "A methodology for assessing sample representativeness", *Environ. Forensics* **6**, 71 (2005). https://doi.org/10.1080/15275920590913877, bit.ly/tos10-5
6. C. Ramsey, "Considerations for inference to decision units", *J. AOAC Int.* **98**, 288 (2015). https://doi.org/10.5740/jaoacint.14-292, bit.ly/tos10-6

7. C. Wagner and K.H. Esbensen, "Theory of Sampling: four critical success factors before analysis", *J. AOAC Int.* **98,** 275 (2015). https://doi.org/10.5740/jaoacint.14-236, 👆 bit.ly/tos10-7
8. K.H. Esbensen and B. Swarbrick, *Multivariate Data Analysis*, 6th Edn. CAMO Software AS, Oslo, Norway (2018). ISBN 978-82-691104-0-1, https://www.amazon.com/Multivariate-Data-Analysis-introduction-Analytical/dp/826911040X/, 👆 bit.ly/tos10-8
9. K.H. Esbensen and P. Paasch-Mortensen, "Process sampling (Theory of Sampling, TOS)—the missing link in process analytical technology (PAT)", in *Process Analytical Technology*, 2nd Edn, Ed by K.A. Bakeev. Wiley, pp. 37–80 (2010). https://doi.org/10.1002/9780470689592.ch3, 👆 bit.ly/tos19-2

11 There are standards— and there is *the* standard

The Theory of Sampling (TOS) is proclaimed to be the only complete theory with which to address all the world's many types of materials with a view to guaranteeing representative samples. This book has made an effort, hopefully appreciated and easy to follow, to explain the basic principles, unit operations and sampling errors and their relationships in this endeavour. However, many standards, guidelines and norm-giving documents (CEN, ISO) already exist, which include *elements* of prescriptions for "proper sampling", such as have been agreed upon by the numerous task forces and committees involved as being appropriate within the relevant scientific, trade and technological contexts. There have been many such attempts towards a general recommended sampling practice, but mostly in a partial sense, indeed *none* cover the full breadth of **all** that is necessary for representative sampling in general (with two exceptions: the ore and mineral processing, and coal industry sectors). With so many partial recommendations, that are demonstrably not in compliance with respect to the TOS, what to do?

AA
ISE: Incorrect Sampling Errors
IDE: Increment Delimitation/Delineation Error
IEE: Increment Extraction Error
IPE: Increment Preparation Error
IWE: Increment Weighing Error
GMO: Genetically Modified Organism
RSV: Relative Sampling Variance
VA: Variographic Analysis
QO: Quality Objectives
FSP: Fundamental Sampling Principle
TSE: Total Sampling Error
SUO: Sampling Unit Operations
GP: Governing Principles

11.1 First light

The publication of DS 3077 represented the world's first standard dedicated exclusively to representative sampling. No other standard is in full compliance with the appropriate TOS requirements laid out here, although *partial* elements can be found in many places, e.g. see the bibliography in DS 3077.[1] Two notable exceptions exist, however: the coal and the iron ore industries, which have been well serviced with excellent standards in this context for many years, in no small measure because of extraordinarily diligent efforts by a very small number of dedicated TOS-expert committee members, most notably Mr Hans Møller and Mr Ralph Holmes.

Non-compliance issues regarding standards, guidelines, good practices as well as regulatory and legal requirements must be handled with insight and patience. Where found not to comply with the TOS' stipulations, it will be necessary to start a process of revision or updating of the relevant standards or norm-giving documents—which may be a lengthy process, and often one that requires quite some logistical and organisational gusto. While this is taking place, or confronted with sampling variances that are demonstrably too high (a key issue in quality control and assurance, QC/QA), it is always an option to employ more stringent sampling quality criteria from a TOS-based approach than what is specified in today's incomplete standards. As there are serious economic and societal consequences of non-representative sampling, simply staying with "following the book" is not a sound strategy, scientifically as well as considering the economic outcome of decisions, which will then be based on inferior, non-representative data.

DS 3077 has the overall objective of establishing a comprehensive *motivation* and *expertise* for taking the stand to rely *only* on TOS-compliant procedures and equipment irrespective of the theoretical, practical, technological, industrial or societal context under the law. No standard is a legal document on its own and is, therefore, not legally binding; all trade agreements ruled by international standards are based on a set of *voluntary* agreements. To the extent that international law on the subjects treated in standards dealing with sampling aspects has been adopted, this law must be adhered to. International law implemented in

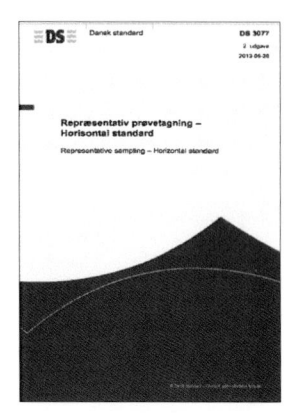

The world's first matrix-independent ("horizontal") standard on representative sampling provides the most concise, comprehensive introduction to the TOS available, powerfully augmented by selected references in this chapter.

national laws also takes precedence to non-legal documents in case of conflict.

Be this as it may legally, there are highly significant advantages in **not** being complacent with the fact that sampling issues are mentioned in the existing body of relevant standards and norm-giving documents. Mentioning is not enough, only the principles *guaranteeing* representativity matter! A directed effort has been in place in the last 15 years, involving a systematic critique of selected standards, specifically with respect to the full set of sampling errors outlined in the TOS. Two examples of this work are presented below, which show how one should approach any part of a standard or similar that *purports* to recommend proper sampling procedures and equipment etc.

11.2 Analysis of sampling standards for solid biofuels

Assessment of all sampling procedures from CEN standards for sampling solid biomass (CEN/TS 14778 part 1 and part 2)[2,3] has shown that most of the recommended procedures do not lead to a satisfactory result: a representative sample. Correct delineation and extraction by many standardised methods as well as depicted, and thus recommended, tools and equipment, are *not* in evidence. While for grab and shovelling methods, correct delineation and extraction is hardly ever possible, other recommended sampling methods lack sufficient specification regarding application conditions, which invariably increases the potential for incorrect sampling error effects. Table 11.1 gives an overview of the evaluation results with respect to potential *incorrect sampling errors* (ISE) caused by the methods stated, recommended or allowed in the standard for primary sampling CEN/TS 14778.[2,3] ISE comprise the four bias-generating errors: Increment Delimitation/Delineation Error (IDE), Increment Extraction Error (IEE),

Table 11.1. Assessment of incorrect sampling errors in CEN/TS 14778.

	Error potential		
	IDE	IEE	IPE
3-dimensional lot	Sampling from stationary lot		
	High	High	Medium
1-dimensional lot	Conveyor belt		
Manual sampling (stopped conveyor belt)	High	Medium	Medium
	Low		
Automatic sampling	High	High	Medium
	Low		
1-dimensional lot	Falling source stream		
Manual sampling	High	High	Medium
Automatic sampling	High	High	Low
	Low	Low	

Increment Preparation Error (IPE) and Increment Weighing Error (IWE), all concerning sampling equipment and sampling procedures. The full assessment of these sampling standards can be found in Wagner & Esbensen.[4]

Insufficient specifications and/or the existence of incorrect sampling errors must under all circumstances be avoided, as the result will unavoidably be an inconstant *sampling bias*, always and forever out of control; it is not possible to make any correction regarding a structural sampling bias.[1,5] Incorrect sampling methods, allowance for personal interpretation and the vertical standardisation approach of specifying different procedures for each material group makes sampling a complicated issue with highly uncertain and varying validity. Any procedure and standard that has not eliminated all such potential sampling bias elements does not comply with the TOS' demands for *sampling correctness*. The result is always a biased sampling procedure. The full assessment and critique of CEN/TS 14778 has been published, but so far no reaction or response has been forthcoming.[4]

11.3   Analysis of grain sampling guide

The "Home Grown Cereals Authority" (HGCA) is a division of the "Agriculture and Horticulture Development Board" based in the UK, which is mainly responsible for research and knowledge transfer in the cereal and oilseed sector. In 2013, the HGCA published a guide on grain sampling to define key requirements for effective grain sampling at various process locations from harvest to storage until departure and arrival of the grain.[6] Besides physical extraction of a grain "sample", focus is on monitoring moisture, temperature, pests and moulds, especially mycotoxins. The described sampling practices must have an obligation to contribute to ensuring that procedures are reliably able to assess harvested grain quality, to protect this quality level throughout the storage phase as well as to determine the quality level after storage (before transportation to buyer), and upon arrival at the buyer. For various commodities the latter two aspects (differences in quality level at departure vs quality level at arrival) have in the past caused major law suits, often due to inappropriate or inadequate sampling procedures. Besides these kinds of discrepancies, which cause serious economic disputes, extraction of representative grain samples is also crucial with regard to impurity detection (e.g. GMO quantification, toxins), as regulated by international standards such as ISO 24276:2006.[7] Table 11.2 gives an overview of the evaluation results for the HGCA[6] with respect to potential TOS-incorrect sampling errors. The full assessment can be found in *TOS forum*.[8]

This assessment shows that most of its recommended sampling procedures and equipment (for both primary sampling and sub-sampling) do *not* lead to a representative sample. The

Table 11.2. Assessment of incorrect sampling errors in the HGCA sampling guide.

Process location (HGCA)	Error potential		
	IDE	IEE	IPE
Sampling at harvest			
Method 1: Sampling before cleaning/drying—Sampling of trailer as it is tipped into store	High	High	Low
Method 2: Sampling after conditioning—Sampling from the cleaner/dryer outlet	High	High	Low
Sampling in store			
Sampling spear (3–5 apertures)	High	Medium / Low	Low
Sampling at outloading			
Sampling from loading bucket	High	High	Low
Automatic bucket sampler	High	High	Low
Sampling from spout loading Jug/Bucket	High	High	Low
Interrupter plate	Medium	Medium	
Sampling from grain heap	High / Medium	Medium / Low	Low
Sampling at commercial intakes			
Manual or automatic sampling spear	High / Medium	Medium / Low	Low

guide's sampling procedures have a high error potential for incorrect sample delineation and extraction, which unavoidably will lead to a significantly detrimental sampling bias.[1] Most of the guide's recommended sampling equipment, when rated with TOS criteria, reveal major incorrect sampling errors (ISE), jeopardising and compromising grain control validity to a varying degree.

It is noteworthy that the body responsible for the HGCA guide undertook a careful response to the above critique, which was published in *TOS forum*.[9]

It is in the general interest of science to bring this kind of discussion and debate to the attention of everybody interested in representative sampling. While the present book does not agree with many of the "reasons for lowering the standard w.r.t. representativity" in the above rebuttal,[9] both science and industry *will* benefit from the clearly stated argumentation vs the original critique. It is, as always, up to the reader to form his/her own conclusions based on the evidence presented *pro et con*.

[9]: D. Bhandari and K. Wildey, "Letter in response to 'A critical assessment of the HGCA grain sampling guide' published *TOS forum* Issue 2", *TOS forum* **4**, 4–4 (2015). https://doi.org/10.1255/tosf.36, bit.ly/tos11-9

11.4 Sampling for GMO risk assessment

Recently, a European Food Safety Agency (EFSA)-funded project engaged in a similar critique of all standards and norm-giving documents governing sampling for GMO risk assessment. The project report is available on the EFSA portal; it can be downloaded from https://efsa.onlinelibrary.wiley.com/doi/epdf/10.2903/sp.efsa.2017.EN-1226, bit.ly/tos11-e

11.5 Examples of too glib recommendations

In the interest of fast learning, this chapter shows a few examples "from undisclosed standards" of "recommended" sampling procedures/equipment, which would not necessarily find acceptance under the systematics of the TOS (Figures 11.1–11.5).

The reader is invited to try to determine which sampling error(s) are committed in each specific example. It is not relevant to refer to the specific standards from which the examples originate; they are shown here in complete anonymity with

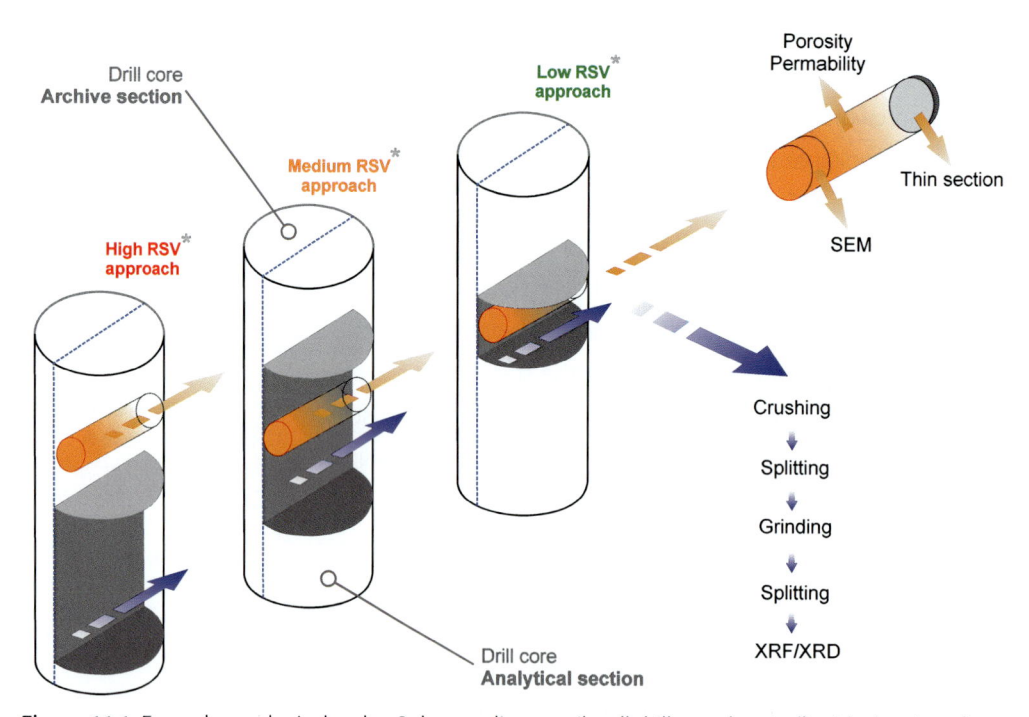

Figure 11.1. From the geological realm. Sub-sampling an oil well drill core (vertical) with the aim of providing samples for diverse analytical modalities [SEM, porosity and permeability measurements (poro-perm), thin section (petrological microscopy), XRF, XRD]. Although a serious attempt at creating a standardised sub-sampling scheme (left: original recommendation), there is room for improvements as seen in the centre and right evolutions towards the smallest sample volume with maximum joint material *support*. There are plenty of arguments from geologists to the tune: "the rocks in the drill core are pretty well homogeneous, so some 15–30 cm vertical separation does not matter much… "; which, however, misses the point. There is no need to create possible IDE and IEE errors in the sampling process, which is tantamount to creating a sampling bias, even though it *could* very well be small—it *could* also be large! Also, homogeneous materials do not exist in the real world, especially not in the poly-phase heterogeneous worlds of rocks… While there may very well be rocks of particular low vertical heterogeneity that need drilling (e.g. sandstone and limestone/chalk oil reservoir rocks), dictating a standardised sub-sampling scheme based on this scenario runs an unnecessary risk of contributing to a possibly significant sampling bias. *RSV: Relative Sampling Variance, see Chapter 9. Reservoir geologists estimate a typical chalk depositional rate of ~2.6 cm per 1000 years (estimates range from 1.5 cm to 5 cm per 1000 years). A vertical sediment interval of, say, 26 cm would thus correspond to 10,000 years. Perhaps a small blink in the eye of geological history, but are reservoir sediments are always laid down on the contemporary sea bottom in a completely identical fashion for some 10,000 years? Better safe than sorry, hence the recommendation above. More information on deposition process can be found in M. Stage, "Recognition of cyclicity in the petrophysical properties of a Maastrichtian pelagic chalk oil field reservoir from the Danish North Sea", *AAPG Bulletin* **85(11)**, 2003–2015 (2001). doi: 10.1306/8626D0D1-173B-11D7-8645000102C1865D, bit.ly/tos11-f

Figure 11.2. A particularly ill-conceived recommendation of a "grain stream sampler". When this example was used as a basis for a university exam question in a PhD course on "Representative Sampling, TOS" a student wrote: "The mind boggles!".

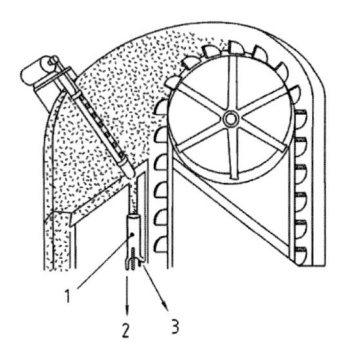

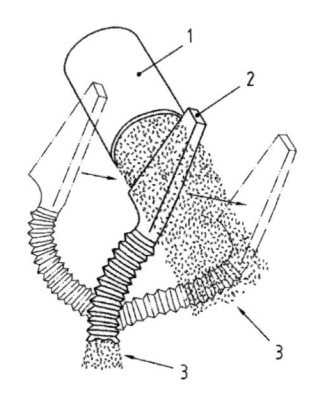

Figure 11.3. A recommended sampler with a much better possibility of being approved by the TOS—why not immediately subject it to a Replication Experiment (RE)? But since the conditions under which this equipment is to operate are totally missing (e.g. transverse cutting speed, cutter opening in relation to the maximum grain diameter), there is still a possibility that it does not live up to all of the TOS' demands.

the sole purpose of illustrating that sampling is not a free-for-all game. More seriously, they are examples of what can happen when committees are guided by a regimen of *consensus decisions* where apparently *anything goes*, as long as it is unanimously voted and agreed on. Pierre Gy often used to deliver a wry comment on this state of affairs in his lectures and courses: "With this approach a committee could vote that Newton's second law no longer applies". The few examples given here are a vivid illustration to this dictum—very many "recommended" sampling procedures and equipment are nothing but a showcase of not having invested the necessary effort to learn the basics of TOS principles. However, there is always room for improvement.

11.6 TOS competence is crucial

There is no need for unnecessary confrontations, but there is a need for absolute clarity with respect to the responsibility carried

> In conclusion, Mr ZZZZZ requested that the committee consider that "The Standard" must support international trade **or it will not be used**.
>
> "The Standard" must be practical and cost-effective.
>
> "The Standard" should be used for both small and medium/large (...fragments...). Other minor commodities could be dealt with in annex.
>
> "The Standard" should take into account differences between continuous and mechanical sampling as well as between manual and static sampling and **indicate a "do the best you can" approach since practical limitations prevent manual sampling across the whole width of the conveyor.**

Figure 11.4. A potpourri of verbatim quotes from discussions in sampling committees and fora. The reader should feel distinctly uncomfortable at the lack of respect for representativity, while logistics, practicality and economics would appear to be the only drivers. The effect of letting such *proxies* dictate sampling procedures, operations and equipment alone was discussed thoroughly, and dismissed by Esbensen and Wagner[5] and Minkkinen *et al.*[12] It is worrying that the TOS competence level of committee members appears to be willy-nilly (to put it mildly); a committee's recommendations are often considered sacred by similarly ill-informed readerships. How, and where, to break this vicious circle? By electing TOS-competent members to the appropriate task forces, technical groups and committees.

Figure 11.5. Using a cylindrical coring tool for cheese sampling (left) does not allow a representative sample of the highly irregularly distributed components of a mature blue cheese. Only the two right-hand approaches will pass muster for the TOS.

by international (and national) standardisation authorities. There is no excuse for recommending non-compliant sampling procedures and equipment; the result can only be inferior sampling and inferior, indeed compromised, decision making. A chain is only as strong as its weakest link. TOS-compliance is the *missing link* in very many standards. There is only one remedy—get involved, get TOS literate!

There are plenty of relevant courses, lectures, consulting companies, experts on the subject matter of representative sampling, all contributing and doing a remarkable job over the last 10–15 years (for some up to 40 years). All the reader needs is a willingness to start looking for the singular operative characteristic: *representativeness*.

11.7   Que faire?

Start here. Below follows an introduction to the origin, history and the contents of DS3077.[1] After having read section 11.8 the following References 1, 5, 10–12 will make up the right progression of more in-depth learning.

11.8   DS 3077 Horizontal—a new standard for representative sampling. Design, history and acknowledgements

July 2013 saw the conclusion of a five-year project, design, development and quality assurance of a new generic sampling standard: DS 3077 Horizontal. DS 3077 Horizontal is published by the Danish Standardisation Authority (DS). Development of this standard was carried out by task force DS F-205. This section summarises the history of this endeavour, focuses on a few salient highlights and pays tribute to the taskforce and to a group of external collaborators responsible for initial proof-of-concept and the final practical quality assurance. DS 3077 describes the minimum Theory of Sampling (TOS) competence basis upon which any sampler must rely in that sampling can be **documentable** as representative, both with respect to accuracy and reproducibility. It represents a consensus based on industry, official regulatory bodies, professionals, university academics and students, and other interested individuals.

11.8.1 Introduction

The primary objective behind DS 3077[1,2] was to develop a fully comprehensive, yet short, easy-to-understand introduction to the minimum principles necessary for sampling all types of materials and lots, at all scales. The overarching goal was to be able to reach absolutely all sampling novices who perhaps earlier had been overwhelmed by the oft-quoted (but wrongly so) impression that the Theory of Sampling is "difficult". This undertaking was ambitious—it took an accumulated 12 core participants in the task force a total of five years to reach a consensus and a product acceptable to all parties. Part of this work necessitated development of partially new didactic approaches, some of which are illustrated below. This section is only allowed to quote a few salient highlights for copyright reasons, but this is enough for an appreciation of the result achieved. The standard has benefitted significantly by valuable input from a large group of external reviewers, assessors, standard writers, sampling consultants and innumerable "users" from science, technology and industry.

Ever since WCSB1, it has dawned upon the international sampling community that there is a serious lacuna in the arsenal with which we try to reach out to *new* communities in science, technology and industry regarding simple, short, easy-to-understand sampling standard. Many attempts had been made but a truly universal standard had not yet seen the light—while very

World Conferences on Sampling and Blending (WCSB)
Readers interested in an introduction to the WCSB conferences are referred to the homepage of the International Pierre Gy Sampling Association: https://intsamp.org/

valuable achievements are on record regarding sampling standards with a restricted target, e.g. for *specific* commodities, major raw materials, manufactured goods etc. These were significant achievements, all of which served as inspiration for DS 3077. A complete "history of DS 3077" can be found in Reference 4. Setting the scene can best be done with a few selected quotes (wider, grey text below), reproduced with permission from the Danish Standardisation Authority, the publisher of DS 3077.

DS 3077 foreword

DS 3077 outlines a practical, self-controlling approach for representative sampling with minimal complexity, based on the Theory of Sampling (TOS). The generic sampling process described and all elements involved are necessary and sufficient for the stated objective, in order to be able to document sampling representativity under the conditions specified. It is always necessary to consider the full pathway from primary sampling to analytical results in order to be able to guarantee a reliable and valid analytical outcome. This standard, including normative and related references, annexes and further, optional references constitute a complete competence basis for this purpose. The present approach will ensure appropriate levels of accuracy and precision for both primary sampling as well as for all sub-sampling procedures and mass-reduction systems at the subsequent laboratory stages before analysis.

A sampling process needs to be structurally correct in order for the essential accuracy requirement to be fulfilled, with no exceptions allowed. For the process also to be sufficiently precise it is often necessary to proceed through iterative stages, until the effective sampling variance has been brought below an *a priori* given threshold; this is also known as 'fit-for-purpose'. In this endeavour the key feature is the heterogeneity of the target lot, which shall be identified and quantified. Heterogeneity characterisation forms one key element of the present standard. Only when both the accuracy and precision demands have been met properly, can all types of solid lots and two-phase (solid–liquid) materials be sampled representatively (gasses are excluded from the present standard), and the derived quality assurance of the sampling process is hereby subject to open public inspection. Without informed commitment to such an empirical heterogeneity characterization, all prospects of being able to document representativity will remain out of reach.

This standard outlines a systematic scientific basis for improving sampling procedures, which will lead to increased reliability for decision-making based on measurement results. Not all existing standards are in compliance with the appropriate TOS requirements, although partial elements can be found in many places (2.1 and Bibliography). Relationships to other standards, guidelines, good practices as well as regulatory and legal requirements shall be handled with insight. Where found in opposition to other, less TOS-compliant stipulations, it will be necessary to start a process of revision or updating of the relevant standards or norm-giving documents which may be a lengthy process. While this is taking place, or when dictated by documented sampling variances that are too high (a key issue in the present standard), it is always an option to employ more stringent quality criteria from a TOS-based approach, than what may be presently codified. As there are serious economic and societal consequences of non-representative sampling, these are appropriately described and illustrated in this standard, which also outlines impacts caused by inferior analytical results and related non-reliable decision making.

DS 3077 has the overall objective to establish a comprehensive motivation and competence for taking the stand relying only on fully TOS-compliant sampling procedures and equipment irrespective of the theoretical, practical, technological, industrial or societal context under the law.

Scope

DS 3077 is based exclusively on the Theory of Sampling (TOS).

DS 3077 is a matrix-independent standard for representative sampling. Compliance with the principles herein ensures that a specific sampling method (procedure) is representative.

DS 3077 sets out a minimum competence basis for reliable planning, performance and assessment of existing, or new sampling procedures with respect to representativity.

DS 3077 invalidates grab sampling and other incorrect sampling operations, by requiring conformance with a universal set of seven governing principles and unit operations.

DS 3077 specifies two simple quality assurance measures regarding:

Sampling of stationary lots, the Relative Sampling Variability test (RSV)

Sampling of dynamic lots, Variographic Analysis (VA), also known as variographic characterisation, with an analogous $RSV_{1\text{-dim}}$. [DS 3077 contains a variographic software program (freeware) making simple variographic characterisation available to all readers]

DS 3077 stipulates maximum threshold levels for both these quality assurance measures.

DS 3077 enforces professional self-control by stipulating mandatory disclosure of one of two comprehensive quality assurance approaches as produced by RSV or variographic characterisation to all parties involved.

DS 3077 specifies documentation and reporting of sampling representativity and efficiency for each analyte in combination with a specific class of materials respectively. Any deviation from this standard's quality objectives (QO) shall be justified and reported.

DS 3077 employs a dual acceptance approach: items not mentioned are not acceptable as modifications in any sampling procedure or sampling plan, unless specifically tested and assessed by the QO's described herein—while all modifications successfully passing this test requirement are acceptable.

This book can only include a small quotation from clause 3 "definitions and terms"; it will suffice here to concentrate on the didactic presentation which has been developed in order to comply with the aspirations re. "short, simple, easy-to-understand ...".

3.11 grab sample
increment resulting from a single sampling operation (literally "grabbing"), almost always emphasizing alleged efficiency, inexpensiveness, effort-minimizing desirability. (Figure 1)

Note: Grab sampling can result in representative samples only in the rarest of instances. If a grab sampling procedure is contemplated, it is mandatory to test and document it by one of the two heterogeneity characterization methods in DS 3077, RSV or variographic characterization.

Grab sampling constitutes the world's most misused sampling operation. All single-sample approaches for heterogeneous materials are in conflict with the Fundamental Sampling Principle (FSP) and militate against the necessary heterogeneity counteraction.

Note: Grab sampling is applicable at all sampling scales, from the field, in the industrial plant to the analytical laboratory, but fails totally to comply with the fundamental sampling principle. DS 3077 mandates composite sampling for all situations in which grab sampling has not been approved by a pertinent validation, either RSV or by variographic analysis."

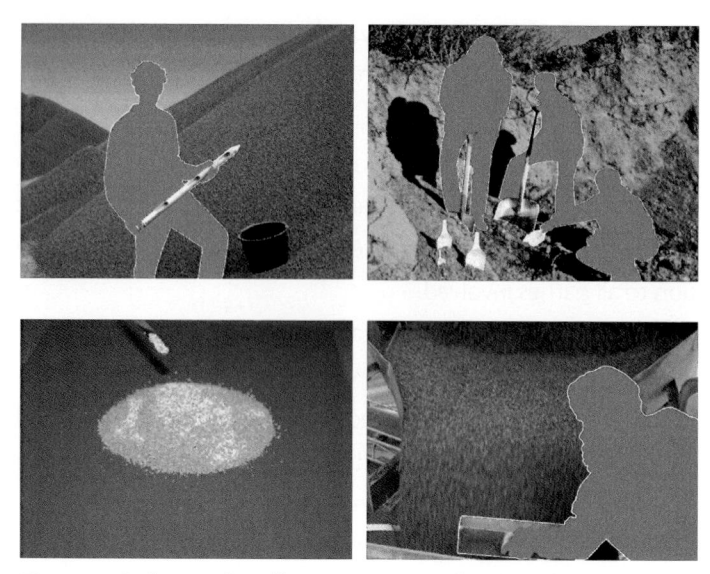

Figure 1. Grab sampling illustration across all scales of interest (from macroscopic stacks to powder piles) for both stationary and dynamic lots. The possibility for any single-increment extraction operation to achieve representativity is virtually zero since the lot cannot be covered with respect to its intrinsic spatial heterogeneity (DH).

3.6 composite sample

sample made up of a number, Q, of increments (Figure 2)

Note 1: The ISO equivalent of a composite sample is the bulk sample. There is full conceptual consistency between the definition of composite (TOS) and bulk sample (ISO), but a composite sample shall either be representative or not, according to the characteristics of how its increments were extracted, a distinction only made in TOS.

Note 2: The primary purpose of composite sampling is to cover spatial and/or compositional heterogeneity of the lot as best possible subject to given logistical and practical conditions and a specific sampling procedure. The same sampling tool (e.g. scoop) can be used significantly better as a provider of a composite sample than when used for grab sampling (single sample operation). In principle, and in practice, informed and competent use of composite sampling will result in a considerably reduced sampling variance (TSE) compared to grab sampling; the average will in general also lie closer to the true lot composition for composite sampling.

Note 3: Composite sampling can also be used for more local purposes, i.e. for minimizing the effect of local heterogeneity (segregation or otherwise) of a single localized sample - for

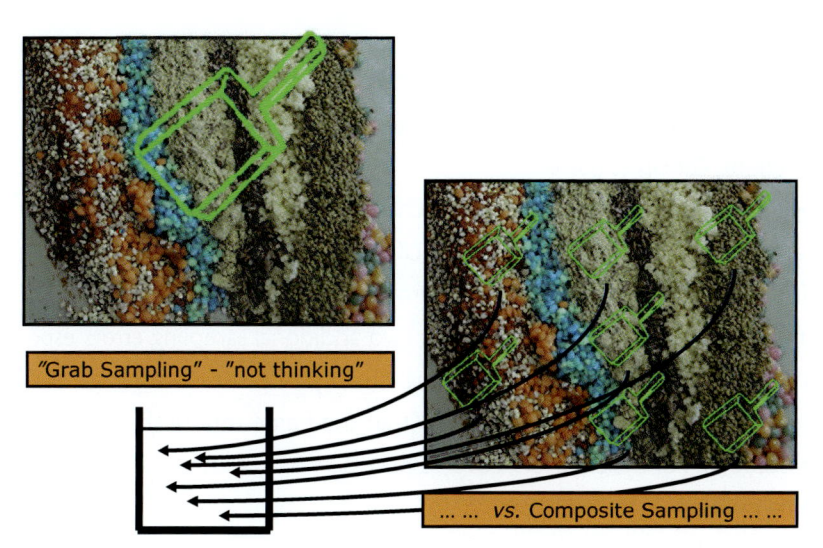

Figure 2. Composite sampling of significantly heterogeneous material. Irrespective of scale, a composite sample (Q increments) is able to "cover" the spatial material/lot heterogeneity far better than a sample originating from a single extraction operation (grab sampling).

example when expressing or modeling concentration changes in 1-D, 2-D or 3-D geometrical contexts, e.g. trend surface analysis."

3.40 theory of sampling, TOS
a body of theoretical work starting in 1950 by the French scientist Pierre Gy, who over a period of 25 years developed a complete theory of heterogeneity, sampling procedures and sampling equipment assessment (design principles, operation and maintenance requirements). TOS was subsequently further elaborated into a coherent didactic framework in the next 25 years by Gy, as well as also added to by newer generations especially in the last two decades. Gy's personal account of TOS and its development history can be found in the note reference immediately below.

NOTE Pierre Gy has published c. 275 papers and seven books on sampling, in later years joined by several other international sampling experts (Pitard, Bongarcon, Minkkinen, Holmes, Lymann, Smith, Carrasco, present author: chairman Taskforce DS F-205). A tribute to Pierre Gy's scientific *oeuvre* can be found in the reference below.[3]

TOS, synoptic overview

The figure below (see Figure 6.4 or the backcover of this book) shows a didactic flow path of relationships between sampling stages, sampling errors, four practical sampling unit operations (SUO) and three Governing Principles (GP).

Empirical heterogeneity testing, RSV (heterogeneity characterisation) is universally applicable, both for the total sampling process as well as for specific sampling stages. Process sampling relies on variographic analysis (VA) for heterogeneity characterization, sample mass (composite sampling, Q) and sampling rate optimization. There are two additional sampling errors especially related to process sampling (trend process sampling error; cyclic process sampling error), which can be brought under control relatively easily. Within the framework of this standard, sampling from either stationary or dynamic lots, covers a necessary basis with which to address very nearly all sampling issues.

11.8.2   Freeware; Variogram

DS 3077 Horizontal contains an appendix comprised by a stand-alone software package, designed to be able to perform basic variographic data analysis for an entry of up to 100 measurements (Figure 11.6). This software calculates a relative variogram on the basis of user input (two spreadsheet columns: concentration, increment weight—if no weight is assigned, the software assumes identical weights for all increments arbitrarily set to 1.00). Variogram calculation is the only option, indeed the only task included. This freeware is in no way intended as a competitor to existing professional and commercial variographic software programs or packages on the market, all of which perform several more essential functions for in-depth usage, e.g. decomposition of variance components originating from periodicity and trends, estimation of TSE. The role of the freeware appendix is *solely* to allow readers an initial familiarisation with variographic data modelling.

11.8.3   Discussion and conclusion

DS 3077 Representative Sampling—Horizontal has been applied on innumerable occasions in the period since its gestation, where it was unanimously concluded that there is a serious need for such a standard. There is no doubt that the present ver. 1.0 is but the beginning on a new journey. As any other international standard it will be subject to regular revision in agreement with the pertinent stipulations (CEN/ISO). It is hoped that many will feel compelled to contribute towards its continuing development and improvement.

11.8.4 Attribution

The taskforce behind DS 3077 (DS F-205) consisted of the following members: Kim Esbensen (chairman), Lars Petersen-Julius, Hans S. Møller, Christian Riber, Anders Larsen, Martin Thau, Jette Bjerre Hansen, Lars K. Gram, Jørgen G. Hansen and Bodil Mose Pedersen. DS 3077 benefitted significantly by valuable corrective and additional input from a large group of invited, external reviewers, assessors, standard writers, sampling consultants and "users" from science, technology and industry. The following individuals are gratefully acknowledged for their major contributions

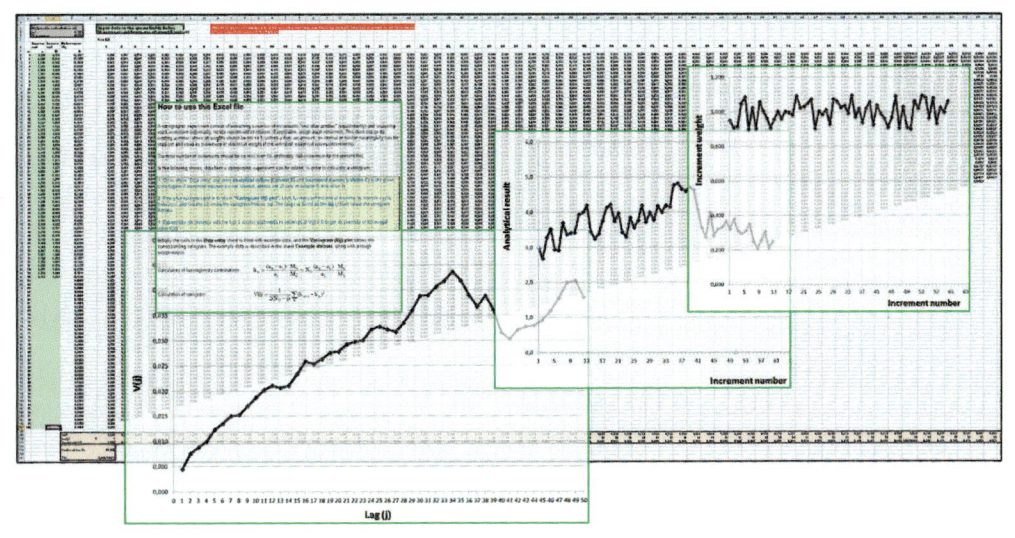

Figure 11.6. DS software VARIOGRAM: Brief program manual, input spreadsheet, plotting routines for concentration and weight values and the resulting variogram (bottom). This software is designed for an input of up to 100 entries.

in this work—but are in no way responsible for perceived errors, omissions or declarative issues in the standard: Francis Pitard, Ralph Holmes, Pentti Minkkinen, Claudia Paoletti, Anna de Juan, Kaj Heydorn, Ulla Oxenboll Lund, Loren Mark, Melissa Gouws, Claas Wagner, Peter Thy, Peter C. Toft, Anders Larsen, Henri Sans and Mark O'Dwyer.

11.8.5   Acknowledgements

Quotes from the "Danish Standardisation Authority", which provides a freely available preview of the first nine pages, are reproduced with permission of Danish Standards.

bit.ly/tos11-d

11.9   Chapter references

1. DS 3077, DS 3077. *Representative sampling—Horizontal Standard*. Danish Standards (2013). www.ds.dk, bit.ly/tos10-2
2. *CEN/TS 14778-1:2005 Solid Biofuels. Sampling. Part 1: Methods for Sampling*. British Standards Institution, London, UK (2006).
3. *CEN/TS 14778-2:2005 Solid Biofuels. Sampling. Methods for Sampling Particulate Material Transported in Lorries*. British Standards Institution, London, UK (2006).
4. C. Wagner and K.H. Esbensen, "A critical review of sampling standards for solid biofuels – Missing contributions from the Theory of Sampling (TOS)", *Renew. Sust. Energ. Rev.* **16**, 504–517 (2012). https://doi.org/10.1016/j.rser.2011.08.016, bit.ly/tos11-4
5. K.H. Esbensen and C. Wagner, "Theory of Sampling (TOS) versus measurement uncertainty (MU)—a call for integration", *Trends Anal. Chem. (TrAC)* **57**, 93–106 (2014). https://doi.org/10.1016/j.trac.2014.02.007, bit.ly/tos1-6
6. *HGCA Grain Sampling Guide*. HGCA Publications, Warwickshire (2013). http://www.hgca.com/media/248889/

grain_sampling_guide_2013.pdf (accessed February 2014), 👆 bit.ly/tos11-6

7. ISO 24276:2006 *Foodstuffs—Methods of Analysis for the Detection of Genetically Modified Organisms and Derived Products—General Requirements and Definition*. International Organization for Standardization (ISO), Geneva, Switzerland (2006).

8. C. Wagner and K. Esbensen, "A critical assessment of the HGCA grain sampling guide", *TOS forum* **Issue 2,** 16–21 (2014). https://doi.org/10.1255/tosf.18, 👆 bit.ly/tos11-8

9. D. Bhandari and K. Wildey, "Letter in response to 'A critical assessment of the HGCA grain sampling guide' published *TOS forum* Issue 2", *TOS forum* **Issue 4,** 4 (2015). https://doi.org/10.1255/tosf.36, 👆 bit.ly/tos11-9

10. K.H. Esbensen, C. Paoletti and P. Minkkinen, "Representative sampling of large kernel lots – I. Theory of Sampling and variographic analysis", *Trends Anal. Chem. (TrAC)* **32,** 154–165 (2012). https://doi.org/10.1016/j.trac.2011.09.008, 👆 bit.ly/tos8-0

11. K.H. Esbensen, C. Paoletti and P. Minkkinen, "Representative sampling of large kernel lots – III. General Considerations on sampling heterogeneous foods", *Trends Anal. Chem. (TrAC)* **32,** 179–184 (2012). https://doi.org/10.1016/j.trac.2011.12.002, 👆 bit.ly/tos8-3

12. P. Minkkinen, K.H. Esbensen and C. Paoletti, "Representative sampling of large kernel lots – II. Application to soybean sampling for GMO control", *Trends Anal. Chem. (TrAC)* **32,** 166–178 (2012). https://doi.org/10.1016/j.trac.2011.12.001, 👆 bit.ly/tos8-2

References DS 3077

1. Danish Standards (DS) (2013): http://webshop.ds.dk/catalog/documents/M278012_attachPV.pdf, bit.ly/tos11-a
2. An authorised preview can be found at: http://webshop.ds.dk/product/M278012/ds-30772013.aspx, bit.ly/tos11-b
3. K.H. Esbensen and P. Minkkinen (Eds), "Special Issue: 50 years of Pierre Gy's Theory of Sampling. Proceedings: First World Conference on Sampling and Blending (WCSB1). Tutorials on Sampling: Theory and Practise", *Chemometr. Intell. Lab.* **74(1),** 1–236 (2004).
4. K.H. Esbensen and L.P. Julius. "DS 3077 Horizontal—a new standard for representative sampling. Design, history and acknowledgements", *TOS Forum* **1,** 19 (2013). https://doi.org/10.1255/tosf.7, bit.ly/tos11-d

12 Spear sampling: a bane at all scales

This chapter focuses on sampling using a popular tool, the "sampling spear" or "sampling thief". There is much good to be said about spear sampling—and only one thing which is bad. But this is bad enough: spear samplers are very, very difficult to get to produce representative samples! The spear sampling principle *can* be made representative, however, in most practical situations in which spear sampling is used today it manifestly is *not*. Why? And more importantly, what can be done about it? This chapter also illustrates one of the TOS' six governing principles: SSI, Sampling Scale Invariance.

A

SSI: Sampling Scale Invariance
REE: Rare Earth Elements
FSP: Fundamental Sampling Principle
DH_{LOT}: Distributional Heterogeneity of the lot
CH_{LOT}: Constitutional Heterogeneity of the lot
TSS: True Spear Sampler
RE: Replication Experiment
ISE: Incorrect Sampling Errors
IDE: Increment Delimitation/Delineation Error
IEE: Increment Extraction Error

12.1 Introduction

It is convenient to begin by presenting a very often used sampler in the laboratory domain, the hand-operated tubular corer (tubular extractor). What is a tubular corer but a (very) small sampling spear designed for forceful insertion into the lot material. This particular sampler is designed so as to allow lot material to be forced into the cylindrical volume as the corer is inserted and forced to greater depths (Figure 12.1).

Earlier in this book it was laid out in detail why the cylindrical corer, used in the one "sample" approach very often is nothing but grab sampling in disguise (we might call this "cylinder grab sampling") if the column extracted does not cover one physical lot dimension completely. The singular cylinder extraction approach is in no way able to produce a representative sample of the highly irregular heterogeneity met with in blue cheese in this

Figure 12.1. Using a cylindrical coring tool for cheese sampling (left) does not allow a representative sample of the highly irregular distribution of components of a mature blue cheese. There is an over-representation of the material towards the centre of the lot. Only the two right-most approaches will pass muster as complying with the TOS' principles of volumetrically balanced increment representativity. For an easy "model" think of a pizza slice (small vertical dimension), or a pie/blue cheese (large vertical dimension); it is both about covering the radial section as well as the third, height dimension appropriately. The reader will easily be able to generalise this model to other 3-D lots where relevant.

example—particularly if the cylinder is applied in the horizontal direction (left photo). If there is directional spatial heterogeneity in a cheese, it is very likely in the vertical direction, even though this is attempted to be compensated for by frequent "turning over" of the maturing cheeses. Even though this standard orientation is aiming at reaching all the way to the centre of the lot (a sound objective), there is a marked volumetric *over-sampling* of the lot material closer to the centre relative to the more peripheral locations (see Figure 12.2).

The illustration of one, or two, *opposing* pie-cuts in Figure 12.1 illustrates the *TOS-correct* delineation of two increments sampling a circular 2-D lot—and takes it further, by expanding the flat lot completely in the third (vertical) direction.

Figure 12.2. Severe over-sampling of the central parts of the lot caused by parallel delineation of the increment; compare with illustration above, Figure 12.1. It does not matter from where this illustration originates—it is wrong! At all scales! For all types of material!

This is a fundamental issue at all scales. Even *if* the tubular corer were of the same thickness as the "cheese" in the third, vertical dimension, it would still be at fault! It would still be *over-sampling* in the central locations (Figure 12.2). The delineating radius vectors must originate at the central vertical axis through the lot, which is not compatible with the geometry of a tubular corer (or the coring tool illustrated in Figure 12.1).

The TOS-correct design of a spear sampler should have been funnel-like, tapering off towards the centre of the lot, but such a geometry violates against a balanced in-flow of material in the corer. Interestingly then (from a TOS perspective), a corer would appear to have to respect two distinctly different geometrical demands, governing vertical vs horizontal insertion. So, the world is left with a plethora of offerings in the form of "universal corers", none of which are able to do correct horizontal coring.

12.2   Spear sampling—at all scales

Spear samplers are popular in all walks of science, technology and industry, and at all scales. Spear samplers range in size from the small scale hand-operated tubular extractors used in laboratories, for example in the food and feed industry, certainly not only for cheese as above, but also for minced or mixed meat products, chocolate, butter and very many, almost innumerable other products. The main purpose is to extract a sample from the *interior* of the lot material, but unfortunately only rarely with a view of getting a *balanced* sample w.r.t. the full lot geometry.

Spear samplers are used extensively also in the meso-scale industrial regimen (1–2 m length) for sampling a wide range of most industrial products and commodities, e.g. grain, fly ash, coal fines, chemical products, construction materials etc. and are furthermore much deployed in bulk materials handling, e.g. for sampling bulk minerals and concentrates, ores, coal, grain, wood shards (biomass and bioenergy sectors). Also, "waste" from other industrial processing that contains valuable elements and compounds that can be recovered at a profit (platinum group metals, Rare Earth Elements (REE), gold, silver etc. ranging in scale from jewellery cuttings to industrial re-cyclates arriving by the truck or railroad load. In many science and technology arenas the characteristics of the target material formally *invites* specific spear sampling, e.g. agricultural and environmental sampling, i.e. of soil and peat or in pharmaceutical manufacturing. This state of affairs

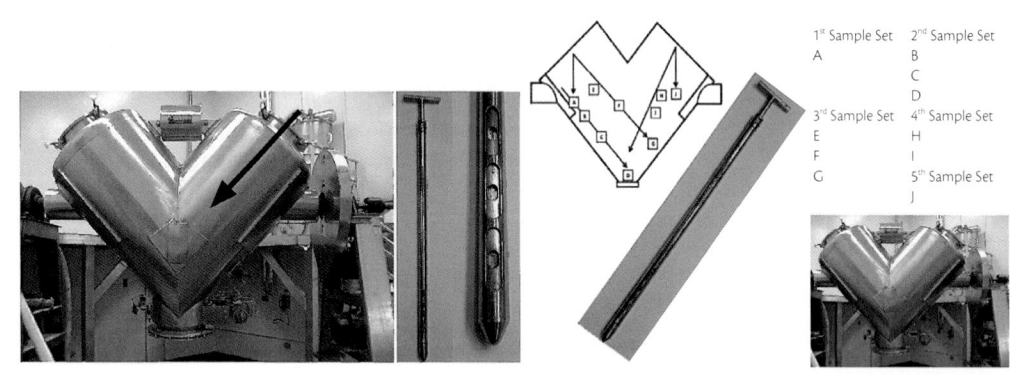

Figure 12.3. Generic spear sampling in the pharmaceutical industry. Although efforts have been made to reduce the increment volume at each designated depth interval (middle photo), identical free inflow of material at progressively larger depths is not necessarily obtainable due to differential compaction with depth. Also (right), it is a fallacy to stipulate that fixed positions within the V-blender (right) are optimal for *all* types of mixtures met with in pharma. The specific pharma spear sampling scenario is described in detail in Esbensen *et al.* (bit.ly/tos12-1),[1] where solutions also honouring the TOS can be found.

is widespread, e.g. spear sampling from big bags, from product bags, from railroad cars, from truck loads..., spear sampling almost *ad infinitum* (Figure 12.3).

All these applications are popular because of the comparative ease with which a column of target material can be extracted. But spear sampling is perhaps mostly popular because of the extremely low capital investment involved, as well as low operator costs. There is actually only one thing wrong with spear samplers—**they are very, very difficult to make representative!**

Against this stands TOS' dictum: representative sampling *must* by necessity comply with the Fundamental Sampling Principle (FSP): all *virtual* increments of a lot must have an identical, non-zero, probability to be extracted, which translates: no physical volume of the lot can be allowed to be potentially out-of-reach of a sampling spear.

From current experience with contemporary practices, it is obvious that *most* spear sampling violates markedly with the FSP demand, because spears only rarely are designed or operated to cover the *full depth* of the lot in question and thus are structurally unable to deal in an appropriate fashion with the distributional heterogeneity of the lot, DH_{LOT}. The crucial issue is to be able to recover, completely and without loss, a *full* core length, and in particular the bottom-most part where absolutely no loss

Figure 12.4. The TOS' Fundamental Sampling Principle (FSP): "All virtual increments must have an identical, non-zero, probability to be extracted". Superficial spear sampling (never penetrating to the inner and core parts of the lot) comprises a severe violation of the FSP.

is allowed—due to segregation. This is the crucial aspect of spear sampling. Violation of this requirement is the most frequent reason that spear sampling is mostly non-representative (Figure 12.4).

If a spear sampler should be able to work in a representative fashion, what might be called a "True Spear Sampler" (TSS), it must by design, manufacturing, usage and maintenance be able to mitigate the deficiencies pointed out:

- For a TSS, the sampling tool must always be able to cover the full depth of the lot (including the "extra" length needed to connect to the driver/engine).
- The TSS must be designed to operate in two modi: forced insertion or true coring (drilling).
- The TSS must be designed always to *recover* the complete core, with special focus on the critical bottom part from which no loss is permitted; this demand is not negotiable.
- The TSS must allow all collected material to be recovered; there must be no material sticking and adhering to the inner surface of the sampler for example.

Any TSS must be tested empirically, under deliberately adverse conditions and with materials comprising at least three components with properties representing mass fluxes and concentrations in typical industrial and technological systems, covering both high, intermediate as well as trace compositional concentrations, see for example, Petersen *et al.* for description of an extensive experimental design of this kind.[2] One of the test components should vary significantly in particle shape, aspect ratio and surface roughness and one of the other should be prone to particle segregation. Such test systems should be as difficult to sample as possible, in order constitute realistic worst case scenarios.[2] Such tests must comply with the stipulations of a proper Replication Experiment (RE).[3] Even if a particular TSS is fit-for-purpose and representative for *some* specific materials, it most emphatically cannot be universally applied to other types of material—*unless* similarly tested empirically by RE. Despite many OEM claims, there is no such thing as a "universal sampler" that will work for all materials… because materials have so dramatically different inherent heterogeneities.

12.3   Not always bad—there is hope

Didactic presentations may sometimes run the risk of focusing too much on what is wrong; this book is probably no exception. But there is much to learn from mistakes. Yet, there is even more to be gained from spectacular success stories which inspire us to think positively, constructively.

So, behold Figures 12.5 and 12.6, which illustrate an innovative spear sampler that **works**—albeit only for free-flowing aggregate matter made up of individual particles. Think of a truckload of grain—this sampler was in fact specifically devised for wheat grain sampling. There is a very large international wheat grain trade by ship, train wagon, lorry truckload and there is an almost infinite demand for reliable QC/QA.

Figure 12.5 shows the principal features of the "RAKOREF" truckload grain sampler as it was originally named by its designers.[a] This is a vertical spear sampler, designed as a truckload corer that pneumatically transfers increments up through an inner cylindrical duct. But most emphatically **not** by forceful suction, which would produce unacceptable in-flow *segregation* (IDE/IEE). Instead, when the spear is pushed into the product, a properly delineated increment is pushed into the inner tubular chamber. This can be seen in Figure 12.6, which shows a close-up of the "zero-pressure" tip

Figure 12.5. The pneumatic grain truckload sampler ("RAKOREF") in all its powerful simplicity; see text and Figure 12.6 for full description.

---

[a]Trade name, originally devised by the now defunct company "Rationel Kornservice (RAKO)", has now been transferred to new owners. Interested readers will be able to find this equipment easily by searching the internet.

concept of the RAKOREF. In a creative solution, no suction effect is applied to the product surrounding the corer, i.e. no lightweight fractions are drawn in from the vicinity of the spear opening (e.g. broken grain particles, dust or dirt). The air flow only works on the increment volume which is present *inside* the spear tip and transports it to a weighing room, or directly to a laboratory, through a PVC hose. There the sampled material is separated from the air, the spear is emptied (often using compressed air) and is ready for immediate use again, while the sample is ready for analysis.

The zero-pressure tubular spear sampler operates identically *as if* it was lowered because of drilling down all the way to the bottom. The brilliant aspect of the RAKOREF principle is that it negates all inflow IDE/IEE effects completely, as long as the downward velocity of the spear tip matches the capacity of the inner cylinder to transport the particular material upwards—an easy engineering task for calibration to any given material. But not only that, this design also dispenses with the potential fatality of not honouring TOS' inflexible demand to extract all the way to, and including, the bottom layer(s), which often are the most segregated variants of the lot material. Indeed as soon as when the tip has reached the bottom, no further influx will take place.

Even though by design confined to one type of material only, one should consider this spectacular solution as an inspiration for how to think out-of-the-box regarding other lots, materials, sampling under other conditions... IF ONLY one could devise a TOS-correct mechanism for closing the spear sampler exactly when the bottom of the lot being drilled is reached—IF ONLY. Which manufacturer will be the first mover?

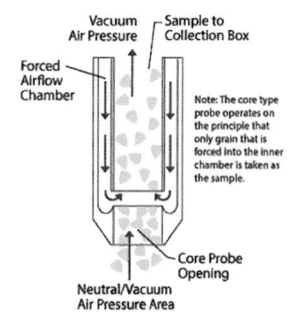

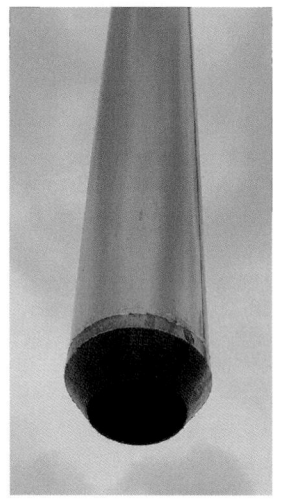

Figure 12.6. The tip of the RAKOREF pneumatic spear sampler allowing "zero pressure" admission of a contiguous series of increments into the inner cylindrical chamber, transporting the composite sample upwards through a flexible hose to a receiving room or to a laboratory.

12.4 Conclusions

Observe how the above functional analysis of "spear sampling" as a generic sampling process is not tied to one or *some* scales only—or to special materials for that matter. The characteristics of spear sampling are principally *identical*—it is only the physical size of the spear sampling tool that changes, so as to match the maximum grain size and the physical lot size.

Note also that lot heterogeneity can change *independently* of the size of the lot and/or the sampling tool. Material heterogeneity is *not* correlated with lot scale, but *is* correlated with, indeed is a function of, the fragment/grain/particle size and the local-scale arrangements thereof (the lot unit elements) contributing to the constitutional heterogeneity of the lot, CH_{LOT}. Thus, proper spear sampling is a function of the unit sampling volume, the increment volume. In composite sampling the increment volume must of course be set so as to *match* CH_{LOT} (influx opening must exceed 3× the largest particle diameter etc.).

The "spear sampler" is a *bad* example of a very often met with misunderstanding: one *type* of sampling tool is declared fit for all purposes, fit for all materials and able to work for all lots … at all scales, which it most emphatically is not!

Yet the spear sampler can be an example of a good engineering solution to a problem that unfortunately is not simple and universal: "how to extract a representative sample from the interior of a lot?", but a problem for which understanding of the full set of concepts and rules in the TOS is necessary, in particular FSP, CH_{LOT}, DH_{LOT}. In order to deal effectively with the latter, DH_{LOT}, it is necessary to understand and acknowledge the imperative of *composite sampling*, i.e. combining a sufficient number, Q, in the form of a complete *stack* of top-to-bottom disc-shaped increments of the lot, see Figure 12.6 and margin box "Exciting developments".

Representative sampling is *not* about buying a *specific* tool with which to take on all the world's materials, i.e. all the world's manifestations of heterogeneity. This is futile. Despite many OEM claims, there is no such thing as a "universal sampler" that will work for all materials… precisely because materials have so many

Exciting developments
Indeed developments in the manufacturing sector are under way. In order to cover a maximum number of materials, careful theoretical analysis, realistic simulations and *extensive* experimental test work is needed. Being able to secure an intact complete column of segregated material (**no** residual material left at the bottom) constitutes the coveted Holy Grail of spear sampling. Examples of significant successes and failures from which to learn during developments are shown here. Illustrations courtesy of HERZOG and KHE Consulting; with permission.

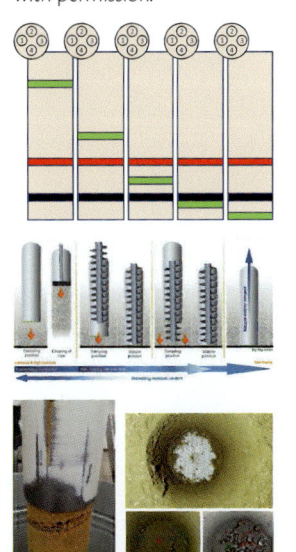

different inherent heterogeneities and other characteristics that profoundly influence sampling in practice, e.g. different moisture levels and, thus, different degrees of clumpiness and stickiness.

Representative sampling is *all* about mastering the necessary and sufficient principles laid down by the TOS[3,4] with which to make rational choices regarding the most appropriate type of sampling tool needed for a *specific* task, for a *specific* material.

Incidentally, the above relates directly to one of the TOS' six governing principles, Sampling Scale Invariance (SSI): when designed, operated (and maintained) correctly (unbiased samplers), the spear sampling *principle* is identical at absolutely all scales.

12.5 References

1. K.H. Esbensen, A.D. Román-Ospino, A. Sanchez and R.J. Romañach, "Adequacy and verifiability of pharmaceutical mixtures and dose units by variographic analysis (Theory of Sampling)—A call for a regulatory paradigm shift", *Int. J. Pharmaceut.* **499**, 156–174 (2016). https://doi.org/10.1016/j.ijpharm.2015.12.038, 👆 bit.ly/tos12-1
2. L. Petersen, C. Dahl and K.H. Esbensen, "Representative mass reduction in sampling—a critical survey of techniques and hardware", in Special Issue: 50 years of Pierre Gy's Theory of Sampling. Proceedings: First World Conference on Sampling and Blending (WCSB1), Ed by K.H. Esbensen and P. Minkkinen, *Chemometr. Intell. Lab. Syst.* **74(1)**, 95–114 (2004). https://doi.org/10.1016/j.chemolab.2004.03.020, 👆 bit.ly/tos12-2
3. *DS 3077. Representative Sampling—Horizontal Standard.* Danish Standards (2013). www.ds.dk, 👆 bit.ly/tos12-3
4. K.H. Esbensen and C. Wagner, "Theory of Sampling (TOS) versus Measurement Uncertainty (MU)—a call for integration", *Trends Anal. Chem. (TrAC)* **57**, 93–106 (2014). https://doi.org/10.1016/j.trac.2014.02.007, 👆 bit.ly/tos12-4

13 Into the laboratory… the TOS still reigns supreme

This book has so far treated primary sampling from lots in all shapes, forms and with nearly all sizes—but the lots treated have all been *larger* than the typical sample on the laboratory bench. The main lesson from the previous 12 chapters was simple and powerful: all types of lots in this size range and all types of materials *can* be sampled based on the exact same principles, codified in the Theory of Sampling (TOS). Differences in lot size, material composition, geometrical form… in one sense do not matter; what does matter, however, is the degree of heterogeneity which has to be *counteracted* by the sampling process. With this and the next chapter we are finally moving into the laboratory, focusing on smaller and smaller lots. It does not matter that occasionally some of these operations will take place in the field (think of a large primary sample conveniently being split down to a significantly smaller size in the field with obvious transportation or other advantages). For systematic convenience we can treat all stages and operations performing sample splitting etc. as **if** they were all taking place **in** the laboratory—without loss of generality.

FSP: Fundamental Sampling Principle
ISE: Incorrect Sampling Errors
IDE: Increment Delimitation/Delineation Error
DH: Distributional Heterogeneity
CH: Constitutional Heterogeneity

13.1 Representative sampling—a scale invariant endeavour

Sampling of small lots of the size typically appearing on the laboratory bench, sub-sampling, sample preparation, final aliquot extraction… all involve an integral element of sampling of the same kind as has taken place before the sample in question arrived in the laboratory (indeed sub-sampling is nothing but sampling at progressively smaller scales). The unifying principles stipulated by the TOS are all with one aim—to make sampling from heterogeneous materials as simple as possible, with the imperative of being representative, at all scales. Thus, one of the most powerful unifying governing principles in the TOS is that representative sampling is *scale invariant*. The physical dimensions of the sampling tools change so as to be commensurate with the particle, increment and lot size. However, the essential issue is that the sampling process at all times (and scales) is 100 % focused on how to counteract heterogeneity (Figures 13.1 and 13.2).

Another of the TOS' simplifying principles is that any properly reduced mass of the original lot can also be viewed as a smaller lot in its own right. This means that at any sampling stage one *may* temporarily view the current lot as a "primary lot" from which to extract a primary sample. This "local viewpoint" is obviously *not* a statement meant to disregard the full pathway from consideration, on the contrary. The local lot is still heterogeneous with all the same ensuing issues… in fact it is *only* the size of the contemporary lot that is different, nothing else. The sampler is facing the *exact* same fundamental problem as when facing the original lot size: how to extract a representative sample from a heterogeneous material? The lot in question just happens to be smaller than its original precursor.

But since the problem is identical, so are the optional solutions: representative sampling is *scale invariant*. It is only the sampling tools that will have to change their physical dimensions—everything else remains identical. There may, or may not, be a smaller distributional heterogeneity in the lot now residing on the laboratory bench, which depends on the preceding sampling process, i.e. whether some sort of *mixing* was carried out along the way. This is actually not an important issue—what is certain

Into the Laboratory... the TOS Still Reigns Supreme

Figure 13.1. The laboratory—the realm of the spatula. At this ultimate stage of the sampling pathway from the original lot, significant sampling error effects can still be incurred, mostly due to a faulty understanding that if the material *appears* homogeneous this justifies grab sampling, i.e. spatula sampling. NOTHING could be further from the truth, however.

Sampling *in the laboratory*: What's the difference w.r.t. the field/plant etc.? *Identical* sampling issues and problems—at all scales …

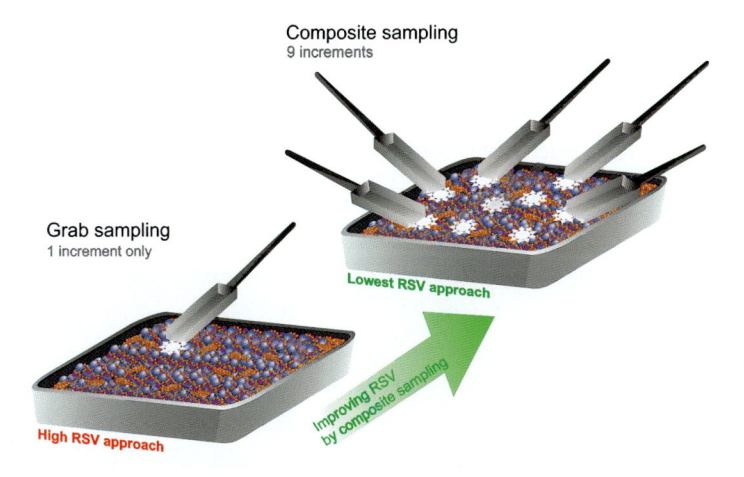

Figure 13.2. First step towards the possibility of a more fit-for-purpose composite sampling procedure in the laboratory. This illustration portrays spatula sub-sampling of a severely heterogeneous material, which decidedly should have been intensively crushed at this stage. Note that even though the FSP is upheld and composite sampling is employed, there are still IDE/IEE associated with the way the individual increments are delineated and extracted, compare Figures 2.1, 8.7 and 13.8.

is that all sub-samples of original heterogeneous lots are also themselves *intrinsically heterogeneous* (mixing can only reduce this heterogeneity, but never eliminate it).

These issues are emphasised with the aim to disallow any argument(s) that sometimes are levelled in order to try to justify that different sampling demands reign at the significantly smaller laboratory scale, or that different types of sampling equipment are needed, or acceptable at this scale. Following the logic of the TOS there can be no other requirements for either procedures or equipment at the laboratory scale than at *all* other, larger scales. The six Governing Principles and the four Sampling Unit Operations recognised by the TOS are the *only* tools available with which to address sampling—at all scales (Figure 13.3).

This understanding has many manifestations and consequences, which at first may appear silly, e.g. "a digger or a front loader, with a one-ton front grabber is *identical* to a spatula"! Identical in the way it may be used *wrongly* to perform grab sampling if only one increment is extracted and *wrongly* declared to be a "sample". It is the faulty grab sampling *procedure* (one increment only) that is identical, albeit performed with radically different sized implements (one ton vs a few grams, perhaps). In this context a digger truly **is** a front loader, —**is** a spoon, —**is** a spatula.

As with all inherent sampling characteristics (governing principles, unit operations, sampling equipment), the task of the competent sampler is to look *behind* the superficial

Figure 13.3. Meso-scale grab sampling (left) vs micro-scale in the lab (right)—what is the difference? This is the wrong question, at the wrong time and at the wrong place (scale)—what matters is that heterogeneity is present all the way to the final aliquot extraction. It truly does not matter whether the human eye can discern material heterogeneity, or not—60+ years of experience allows the TOS to state categorically that *all* materials are significantly heterogeneous and should, therefore, *always* be treated accordingly. This insight actually makes sampling immensely easy: act as if *all* lots, *all* materials, at *all* scales are *always* significantly heterogeneous.

Figure 13.4. Sampling tool size should be set to match the lot and average grain size, but the key function of any sampling tool is always to perform the increment extraction in a way that promotes the counteraction of the material heterogeneity present (CH, DH), i.e. to perform composite sampling.

manifestations, to find out whether current sampling usage actually comply with the simple demands of the TOS, or not. If not, it will never be possible to qualify a particular sampling process as representative, no matter how "ingenious", "smart", "labour-saving", "practical"… at first sight. Manufacturers' websites and marketing brochures, various homemade industrial "solutions", and the literature are nevertheless crammed with such offerings. Later in this introduction to the TOS, a follow-up Book II will contain chapters focusing on such misguided information, termed "Sampling, Hall of Fame" and "Sampling, Hall of Shame". See Figure 25.2 for a particularly illustrative example.

Figure 25.2. A representative sampler? This example is discussed in full in Chapter 25.

13.2 Size does not matter—only heterogeneity, and how to counteract it

Figures 13.5–13.7 give a few illustrations, all from the laboratory realm. Observe how there is absolutely no difference here with respect to examples presented in earlier chapters relating to larger scales. Once you have understood the simplicity of the governing principle "representative sampling is scale-invariant", you will experience a massive empowerment. Lot and sample/increment size never matters again.

It is fair to state, however, that this insight has not always been fully comprehended in the gamut of scientific, technological and practical sampling literature. There are very many examples with which to justify this statement, but only a few spectacularly illustrative cases need to be presented here. Figure 13.6 points forcefully to the issue that there must always be a *unified* sampling responsibility all the way "from-lot-to-aliquot", of which more later.

13.3 And there is more to be done in the lab ...

Other chapters detail the complete systematics of proper TOS application at each sampling stage. Here, it is emphasised that it

Figure 13.5. Even at laboratory scales, segregation may present serious heterogeneity problems. Composite sampling is imperative, with the critical proviso that all increments must cover (counteract) heterogeneity in the vertical direction (see previous chapter on "spear sampling"). The exact same principles apply in the laboratory as everywhere else.

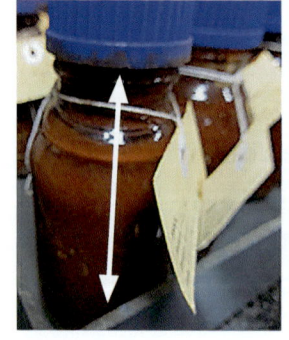

Into the Laboratory... the TOS Still Reigns Supreme 157

Figure 13.6. Perhaps the world's most misplaced sub-sampling call: in the process of crushing carefully collected 12 kg composite field samples (>16 increments, as illustrated), assisting students were told by a laboratory head to "forget all this TOS stuff—the **usual** procedure here is to select a lump the size of what is needed for further treatment (further fine crushing) and only crush this mass instead of all the silly 12 kg" (the indicated lump is circled—20 g). Luckily, the students involved were sufficiently competent w.r.t. the TOS to have the courage to neglect such "advice". There was a quite specific reason why the field composite samples weighed in at a minimum of 12 kg—specifically to counteract the troublesome field heterogeneity encountered. The suggestion to skip the crucial full laboratory crushing stage would have produced a ~600 times smaller sub-sample (12,000 g / 20 g), essentially grab sampling at this sub-sampling stage, which would have reduced the primary composite sampling objective to a completely uncontrollable degree. There is also a critical need for appropriate TOS knowledge in the analytical laboratory: from the bottom up, i.e. for technicians, as well as supervisory personal and indeed managers as well.

always helps towards major reduction in sub-sampling variability to apply the unit operation crushing/grinding/comminution. Whenever grinding is an available option, *always* to be followed by efficient mixing, substantial improvements will always result. A systematic outline of these advantages can be found in the

158　　Introduction to the Theory and Practice of Sampling

Figure 13.7. A show of futility. Shown here are typical sampling and sub-sampling tools in a professional analytical laboratory. The issue was which tool is optimal for final analytical aliquot extraction: spoon vs spatula (the fork is supposed to be a mixing implement)? Grab sampling reigned supreme even at this ultimate, smallest scale. Note the potential spatula replacement in the background, the "Ingamells' micro-splitter".

Figure 13.8. Illustration of an effective laboratory sub-sampling approach, the "Japanese slab cake", from which individual increments can be extracted (at random, or from a regular grid as shown). This illustration also shows how even the last spatula step can in fact still give rise to both IDE and IEE effects (top panel), compare with Figure 2.1. TOS diligence is needed—all the way to analysis!

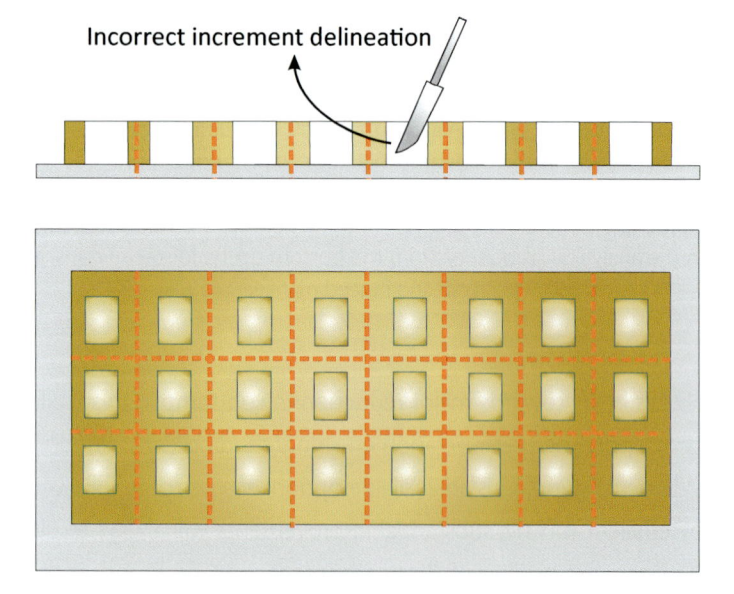

comprehensive paper by Dubé et al. (2015), which, although strictly speaking only addresses "soils", has significant relevance to very many other types of materials. This paper should be considered mandatory for all readers: it adds much to the introductions in the present chapter.

Figures 13.7 and 13.8 illustrate yet more possibilities for doing the wrong thing even with the right intent. One can even use a spatula in a fashion producing IDE/IEE... who would have thought?

13.4 Further reading

P. Bedard, K.H. Esbensen and S.-J. Barnes, "Empirical approach for estimating reference material heterogeneity and sample minimum test portion mass for 'nuggety' precious metals (Au, Pd, Ir, Pt, Ru)", *Anal. Chem.* **88**, 3504–3511 (2016). https://doi.org/10.1021/acs.analchem.5b03574, ☛ bit.ly/tos13-1

J.-S. Dubé, J.-P. Boudreault, R. Bost, M. Sona, F. Duhaime and Y. Éthier, "Representativeness of laboratory sampling procedures for the analysis of trace metals in soil", *Environ. Sci. Pollut. Res.* **22(15)**, 11862–11876 (2015). https://doi.org/10.1007/s11356-015-4447-1, ☛ bit.ly/tos13-2

K.H. Esbensen, "Materials properties: heterogeneity and appropriate sampling modes", *J. AOAC Int.* **98**, 269–274 (2015). https://doi.org/10.5740/jaoacint.14-234, ☛ bit.ly/tos13-3

K.H. Esbensen and C. Wagner, "Theory of sampling (TOS) versus measurement uncertainty (MU) – a call for integration", *Trends Anal. Chem.* **57**, 93–106 (2014). https://doi.org/10.1016/j.trac.2014.02.007, ☛ bit.ly/tos12-4

E. De Andrade, K.H. Esbensen, M.C. Fernandes and A.M. Lopes, "Amostragem representative para uma quantificao precisa de sementes geneticamente modificadas", Ed by P.S.

Coelha and P. Reis. Agrorrural – Contributos Cientificos, pp. 604–615 (2011). ISBN 978-972-27-2022-9

L. Petersen, C. Dahl and K.H. Esbensen, "Representative mass reduction in sampling – a critical survey of techniques and hardware", in *Special Issue: 50 years of Pierre Gy's Theory of Sampling. Proceedings: First World Conference on Sampling and Blending (WCSB1)*, Ed by K.H. Esbensen and P. Minkkinen, *Chemometr. Intell. Lab. Sys.* **74**, 95–114 (2004). https://doi.org/10.1016/j.chemolab.2004.03.020, bit.ly/tos12-2

14 Representative mass reduction in the laboratory: riffle splitting galore

While this introductory book presents the universal principles behind representative sampling of all types of lots and composition, the focus has so far mainly been kept outside the analytical laboratory. This is because many are of the opinion that applying the Theory of Sampling (TOS) at all such large(r) scales (primary sampling) is different from the work that is thought to be specific for the analytical realm, which indeed takes place at much smaller scales—"within the four walls of the laboratory". However, if the systematics of the TOS are to be used to their full power and reach, this division needs careful attention—it is time to enter the lab.

AA
IEE: Incorrect Extraction Error
IDE: Increment Delimitation/Delineation Error
ISE: Incorrect Sampling Errors
CSE: Correct Sampling Errors
GLP: Good Laboratory Practice
SUO: Sampling Unit Operations
TSE: Total Sampling Error
RCS: Rotating Cone Sampler
RSS: Rotary Stream Splitter
vCSS: Variable Cross-Stream Splitter

14.1 Introduction

Truth be told, some of the many operations falling under the term "sample processing" or "sample preparation", and some elements of "sample presentation" correspond to straight-forward sampling processes—only *writ small*. As is shown below, it pays well to follow the TOS' universal application scope all the way to its ultimate goal, that of selecting (sampling for) the analytical aliquot. It is highly advantageous to view absolutely **all** sampling operations, spanning the entire "from lot-to-analysis" pathway, as a scale-invariant theatre in which the sampling operations are in principle identical. Thus a spatula – is a laboratory spoon – is a shovel – is a spade – is a backhoe grabber – is a crane grabber... All these tools are used to select and extricate an increment, it is only *the scale* that varies. The choice of which sampling tool dimension to choose is primarily related to the lot size vs the desired increment size, all of which is also related to the grain size characteristics of the lot. Collecting an increment, or several, is in practice always related only to two possible options: to perform grab sampling **or** composite sampling. Earlier chapters dealt with these systematics in detail, in particular from the perspective of a particularly popular tool, the sampling spear, aka the sampling thief. The present chapter complements this, and deals exclusively with the most often used method for mass reduction approach in the lab—riffle splitting.

14.2   Riffle splitting

There are only a few requirements needed in order for riffle splitting to be correct and effective, by which is meant the most effective way to obtain representative mass reduction in the lab. The sample material must be free-flowing in order to be able to pass through the riffles, driven by gravity. Other than that, there are obvious requirements related to the largest particle size (in some less frequent cases also related to the sorting of the material). In general, it is obvious whether a specific target material is suitable for riffle splitting or not. It is the largest particle size that determines the operative requirements of the riffle splitters. A well-known rule of thumb is that the individual riffle opening must be three times the largest particle diameter + ε,[a] in order to prevent all possibilities of *clogging* a riffle chute. With these few

---

[a] Be advised that in some sectors of technology and industry, for some reason, a tradition has developed to state that this demand is 2.5× the largest diameter. This is logically and factually wrong, however, and is to be discontinued with extreme prejudice!

Representative Mass Reduction in the Laboratory

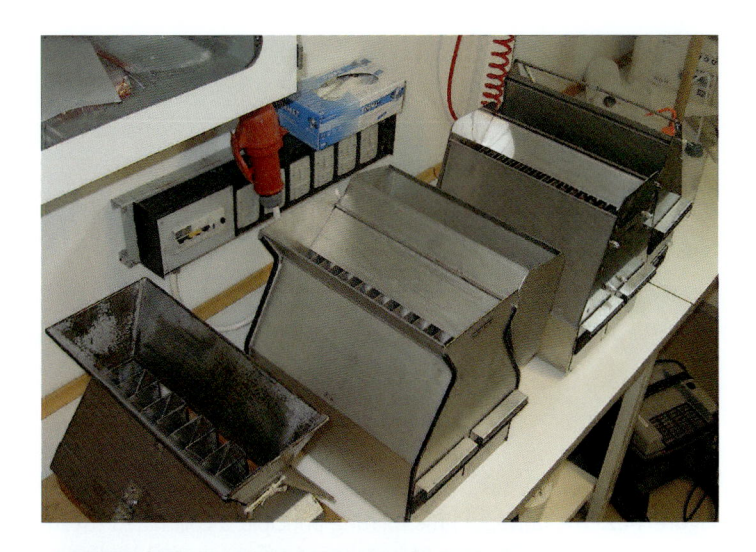

Figure 14.1. Size does not matter much. Riffle splitters are available in a large range of sizes, determined by the effective opening of the riffles (chutes). The smallest met with so far is illustrated on the far-right lower panel, managing to compress 14 juxtaposed chutes along a linear distance of only 5 cm. The resulting chute width is just about the smallest opening that can accommodate very fine grained aggregate material and powders without a serious danger of clogging (and only for materials with relevant characteristics). Riffle splitters with openings smaller than this do not exist, and other ways must be found (other tools) that manage to do sub-sampling in a fashion that achieves the same purpose.

requirements in place, riffle splitting is largely *scale-invariant* and one may pick the splitter tool that fits the practical and logistical conditions and requirements best, see Figure 14.1.

The riffle splitting principle can be implemented in a great variety of ways, and is realised by a wide range of tool brands, but the principle behind it is intuitively simple and easy to comprehend: the objective is to split an incoming mass into two *equal* sub-samples both with respect to mass but also (which is decidedly most important) with respect to the analyte concentration

Figure 14.2. The universal riffle splitting principle: a collimated stream of matter is split by a series of juxtaposed riffles (chutes) leading to a number of slices of the stream being collected into two alternative sub-sample reservoirs. The similarity to composite sampling is striking: in fact iterated riffle splitters are sometimes expressly used for mixing purposes, with appropriate collection in one, or both, sub-sample reservoirs. Illustration courtesy Hans-Henrik Friis-Pedersen, with permission.

to be found in each split. There are also variations aimed at different splitting ratios, see further below. The universal riffle splitting principle is illustrated in Figure 14.2.

Perhaps surprising, it is fully possible to conduct riffle splitting in a non-representative manner. Thus, there are rules governing riffle splitting if this is to be representative. Some of these are illustrated below as a first foray into the subject.

👉 L. Petersen, C. Dahl and K.H. Esbensen, https://doi.org/10.1016/j.chemolab.2004.03.020, bit.ly/tos14-a

> For complete coverage of this critically important subject, see Petersen et al.[1] or Pitard,[2] but these basic issues are also covered in many of the background TOS literature references, see, for example, in the standard DS 3077.[3]

There is always a danger that *some* of the component particles may accidentally bounce or rebound upon hitting the chute separator edges a.o. and thus be propelled out of the active chute

Representative Mass Reduction in the Laboratory

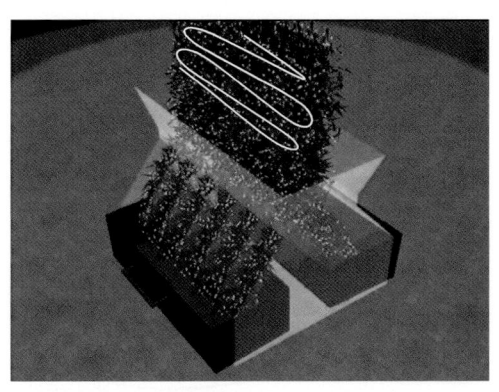

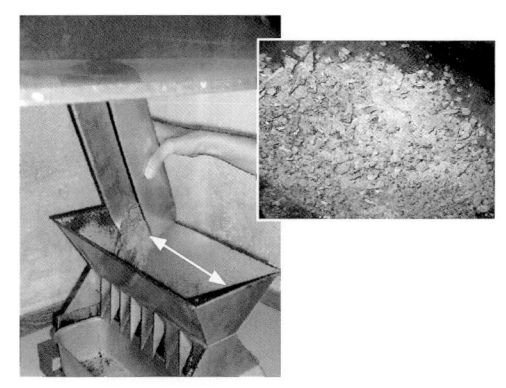

Figure 14.3. Longitudinal "to-and-fro" loading of the ingoing sample to be split is often an area of major misunderstanding. This "covering all chutes evenly" operation may well seem fair and reasonable at first sight, but, upon scrutiny, it is revealed to be based on a faulty, unjustified assumption that the material in the loading tray is fully homogeneous. As is very well known (see all previous chapters), this never occurs in the world of science, technology and industry, and will always result in an unnecessarily inflated sampling error at this stage. The loading approaches shown here should be avoided; see the following illustrations for correct operation of riffle-splitters.

splitting zone. Such components are *lost* from the splitting products, i.e. an Incorrect Extraction Error (IEE) has been committed (by a structural Incorrect Delineation Error, IDE). Figure 14.4 illustrates why riffle splitters always must be *closed* or encapsulated. This is not too much to demand from any manufacturer.

In the right-hand image of Figure 14.4, a serious effort has been made to prepare the loading tray so as to deliver all the material along the longitudinal splitter axis in a controlled, even fashion (see also Figure 14.5). This, combined with the preceding effort to mix the material in the tray thoroughly before loading, results in subsequent splitting procedures with a significant reduction in both Incorrect as well as Correct Sampling Errors (ISE, CSE).

The above precautions can always be observed, it is only a matter of the TOS joining forces with Good Laboratory Practice (GLP). So, many repetitive mass reduction operations can be

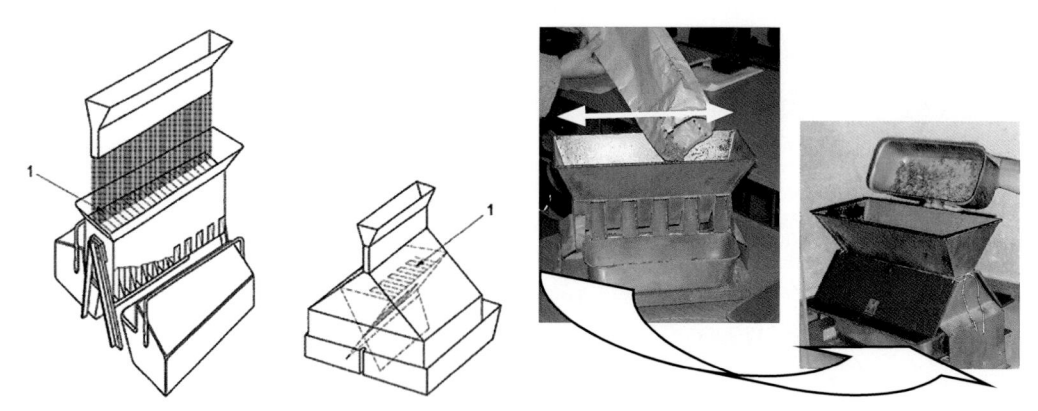

Figure 14.4. Left: how the closed equipment requirement can be easily realised, or not. Right: how the misunderstood "covering all chutes" *to-and-fro* loading is replaced by a carefully prepared loading tray being used so as to deliver all the material along the longitudinal splitter axis covering all chutes *simultaneously* in a controlled, even fashion. There has also been a serious effort to mix the material in the tray thoroughly before loading. All such operations help! Also note (right-most photo) how a very inexpensive, simple way of closing the splitter is a very effective guard against dust emission (dust spillage).

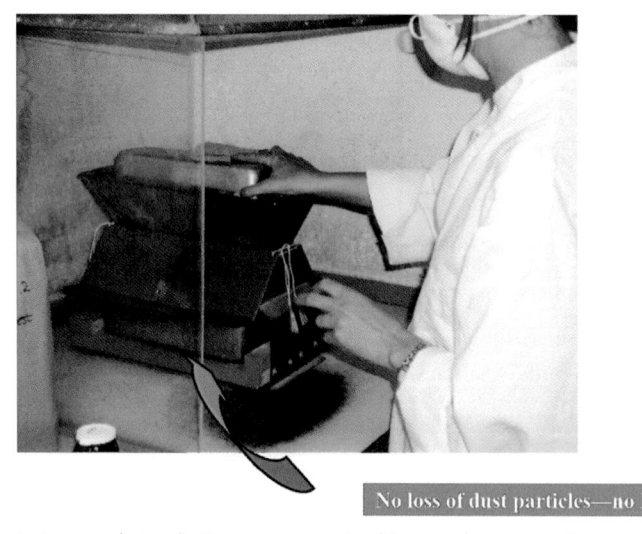

Figure 14.5. Albeit using only primitive and simplistic prototypes, in this sample preparation room, en suite the analytical laboratory, the riffle splitting operations shown do everything correctly, indeed in a representative fashion: no IDE, IEE because of correct pre-loading mixing, correct loading, using a correct enclosed splitter, the latter feature eliminating the otherwise crippling IEE due to dust spillage. With but a little care, all sampling bias dangers can be avoided everywhere after the primary sampling stage.

carried out, across quite extensive scale ranges, with only small efforts (there is always relevant equipment to be had).

As an example with wide application: more thorough crushing is a very effective sampling unit operation (SUO) that can be used with remarkable effect. Figure 14.6 shows how a modestly improved comminution results in a significantly improved loading tray material constitution, much better suited for improved splitting efficiency. Dictum: one *always* mixes well after **any** crushing operation!

It is in the *interaction* between optimised material constitution (mixing and evenly loaded) and the number of riffle chutes brought to bear, where riffle splitting mass reduction really comes to the fore—with a significantly improved (i.e. reduced) TSE. In Figure 14.7, a very heterogeneous, very unevenly distributed material (here deliberately almost a caricature) in the loading tray is subjected to a series of different splitting chutes (8, 16, 32), making it obvious that an increased

Figure 14.6. More efficient crushing leads to a much more uniform material in the loading tray, especially when combined with a conscious effort for better mixing and spreading out over the full loading tray bottom, as shown in the top right-hand photo, excellent TOS GLP!

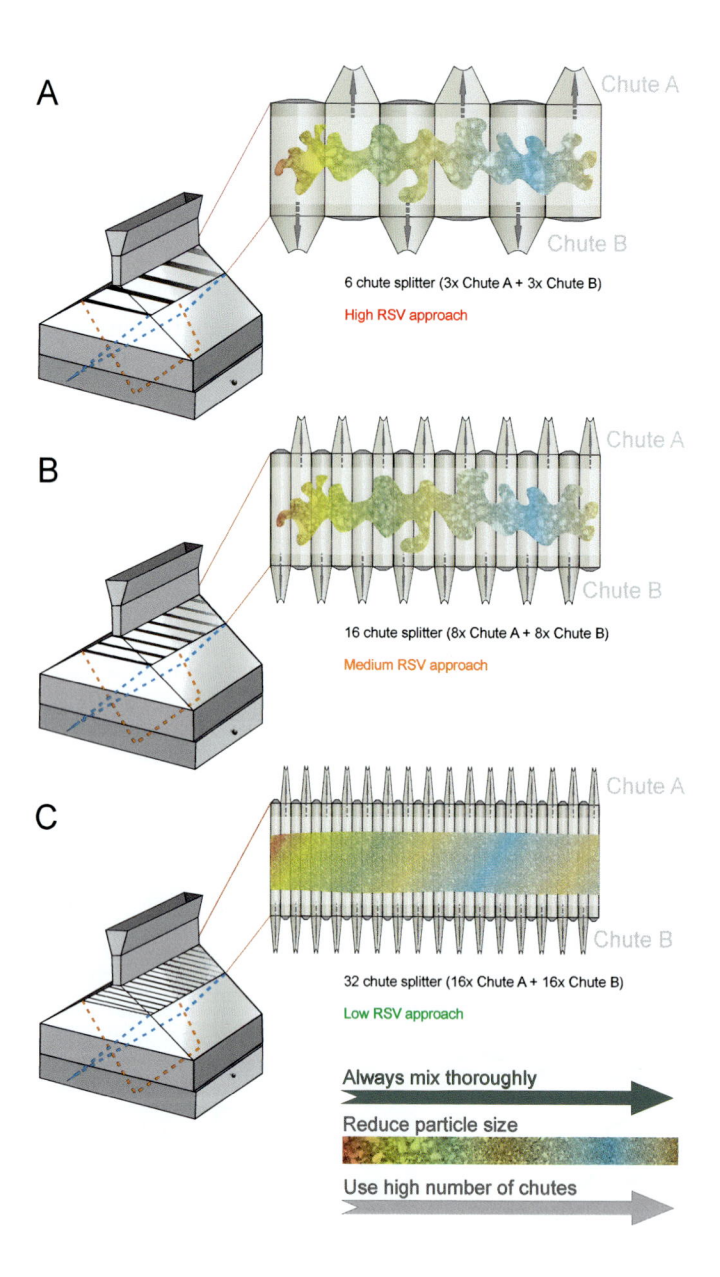

168 Introduction to the Theory and Practice of Sampling

Figure 14.7. Comprehensive illustration of an unequalled TOS-guided GLP approach to optimal riffle-splitting that makes use of progressive crushing and increased mixing in the loading tray before being subjected to an increasing number of chutes. The more crushing, the more mixing, the higher number of chutes—the better, i.e. the smaller laboratory TSE. Illustration courtesy Mr Martin Lischka, with permission.

number of chutes always offers better sub-sampling, everything else being equal. Figure 14.7 shows the situation in which the most effective splitter (32 chutes) is brought to bear on a much improved material constitution (much better crushed and very well mixed material). The essential issue here is that it is the exact same material subjected to very different riffle splitting operations. There is no doubt that when crushing, mixing and effective riffle splitting are brought together with a well-considered, TOS-informed plan for the material's characteristics, the largest reduction in TSE can be obtained at essentially no extra effort.

Observe how proper riffle splitting (using, say, G chutes) acts like a very thorough composite sampler—each of the two identical sub-samples were constructed by aggregating $G/2$ increments covering the entire lot (the incoming load sample). This is why pre-splitting mixing and making sure that this materal is spread out in the loading tool as evenly as possible is of paramount importance. This composite sampling effect is not always recognised.

Thus, Figure 14.7 illustrates the significant advantages obtainable when calling in three of the four Sampling Unit Operations (SUO) in their right *order* (crushing, mixing, composite sampling) leading to the most efficient (least TSE) mass reduction possible in the analytical laboratory. Compare this to the plethora of sub-optimal, indeed often fatal, applications of grab sampling which can be observed in many of the world's laboratories in which the grab sampling scoop or the spatula approach still rules.

14.3   Automation—enter the rotary divider

At one time or other the advantages of using riffle splitters for effective TOS-correct mass reduction will be(come) obvious. All the necessary, but repetitive, manual work will at first be a blessing because of the dramatically reduced TSE involved. Soon, however, *all* this work will begin to look like a burden—"if only this work could be *automated…*".

Well, no problem: enter the *rotary divider*. Rotary dividers act and function precisely *like* a riffle splitter, in fact they *are* riffle splitters through and through, only designed for a much more efficient *throughput*. Figure 14.8 shows two versions of the rotary divider, one with fixed opening widths for the number of chutes chosen (32), and one with a variable chute width for the number of chutes chosen. For both there are now no limitations as to the weight of the sample to be loaded, because any (large) sample mass can be loaded into the hopper in successive parts without changing the sum-total splitting operation; this is a huge advantage both for the high-throughput laboratory as well as with respect to on-line process implementation. Both the rotary dividers shown here operate on the basis of the same framework, with a loading hopper and a rotating nozzle that delivers a steady stream of material hitting the splitting chutes which are arranged in a circular fashion ("radial chutes").

By carefully balancing the loading flux in relation to the rotating nozzle speed it is an easy task to arrange for the sample mass to be split and distributed over a very large number of chutes;

Figure 14.8. The rotary divider—the ultimate mass reduction equipment. Many variants of this principal solution can also be automated. A comprehensive benchmark study can be found in Reference 1.

every new 360° turn of the nozzle allows the stream flux to be distributed over a new multiple of the fixed number of chutes (32, 64, 96 …); the number of operative chutes multiplication factor is staggering, making rotary dividers very much more efficient compared to their stationary, linear cousins. There are many other advantages associated with rotary dividers, see References 1–3, not least that they can be implemented in a continuous flow situation, or can also function as an on-line by-pass stream divider. This latter has many advantages, especially as this makes the splitter insensitive to fluctuations in the stream flow. Such samplers are called proportional samplers.

Proportional samplers, see http://bit.ly/tos14-4

Proportional sampling is an advanced development for sampling of process streams with variable flux (load). For the interested reader a thorough introduction can be found in Reference 4.

14.4 Benchmark study

An almost infinite set of variations on the theme of laboratory mass reduction approaches and methods exist, which type of equipment to use etc. Upon scrutiny and reflection, however, there are only a limited number of *types* of procedural approaches: grab sampling (spatula, spoon etc.), riffle splitting (linear, rotary), coning-and-quartering, asymmetric in-line dividers … A little systematics will give the reader a comprehensive overview.

Figure 14.9 presents a graphical synopsis of the gamut of what is being used today in science, technology, industry and commercial laboratories—starting with grab sampling, i.e. using *one* extraction to get the analytical mass directly (TOS: obviously fatally wrong if/when homogeneity has not been documented beyond reasonable doubt), through "shovelling methods" with various fractional shovelling ratios (akin to simplistic composite sampling), to the well-known "spoon method" (used extensively in the seed industry) and the "Boerner divider" (a well-nigh brilliant invention from the same realm, Figure 14.10), to linear as well as rotary riffle dividers.

And then there is *coning-and-quartering*, which turns out to be the world's most misunderstood combination of inferior mixing followed by a type of extremely coarse four-riffle splitting—to be avoided at all costs. Coning-and-quartering (C&Q) was treated in detail in a paper that *could* have had the title: "Why we killed C&Q and why it had it coming", but which was supplied with a more scientifically acceptable title (with the exact same meaning, however), see Reference 5 for the full story.

At the other extreme is the "Boerner divider". The functional principle is gravity driven, equal azimuthal cone-dispersion,

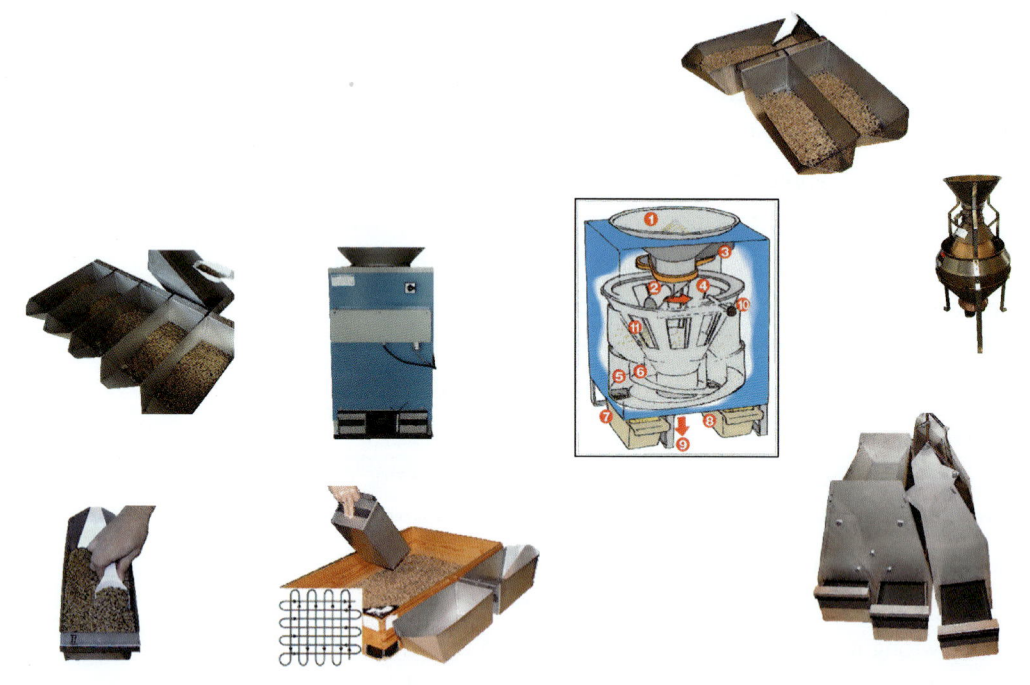

Figure 14.9. Overview of the principally different mass reduction methods and typical types of equipment in the authoritative benchmark study by Petersen *et al.*[1]

Figure 14.10. The famous "Boerner divider", functioning exactly like a rotary divider but without moving parts. Every second chute leads to two separate collecting circumferential funnels (inner and outer), allowing complete separation into two as closely to identical splits as possible (identical sub-samples).

sectorial chute splitting (34 chutes) *without* any moving parts. The principle is sheer genius, and is illustrated well in Figure 14.10. Better still: "Look it up, look it up—*Google it*". Not only is the design brilliant, its appearance is often also a thing of beauty (such balanced use of brass and copper).

14.5 The ultimate method/equipment ranking for the laboratory

The present introduction to the typical objectives, methods, equipment design and means-of-operation for mass reduction in the laboratory has been swift, but hopefully comprehensive and inspirational enough for the reader to wish to follow up immediately—start with Reference 1. In fact, all the principal types of mass reduction methods used in today's laboratories in science,

Mixing: the often forgotten criticality
An underlying theme behind most of the performance of the sub-sampling and splitting operations in the laboratory is the degree of mixing obtained. For the interested reader, a comprehensive introduction to this topic can be found in References 7 and 8, each of which are an eye-opening tour de force for all laboratory personnel.

technology and industry are covered there, as are their typical practical manifestations.

Figure 14.11 is the summary *representativity ranking* of all methods and approaches covered in Reference 1. From the TOS' definition of representativity: $r^2 = (\text{bias})^2 + (\text{imprecision})^2$; *the smaller the r^2 the better the sampling*, i.e. the splitting approach/method/equipment. Detailed scrutiny of the plot reveals the general conclusions of this extensive benchmark:

- Shovelling methods of all kinds are *unacceptable* (excessive TSE, excessive r^2);
- The riffle splitting *principle* reigns supreme, rotary over linear when possible, but both variants work exceedingly well—critically dependent on proper eradication/reduction of all ISE, CSE);
- The "Boerner divider" is superior to pretty much anything else. The gravity-driven symmetrical cone splitting principle has recently been further developed to construct a "Rotating Cone Sampler" (RCS), which in an ingenious manner makes full use of this principle for the purpose of in-line implementation.[6] First performance surveys are generally encouraging, but also reveal intricate material-dependent issues in need of further refining.[6]

There also exist recently designed but still "classical" dividers of various types, specifically to service in-line splitting in the automated laboratory context. Two such are the rotary stream splitter (RSS) and the variable cross-stream splitter (vCSS), which are described in Reference 7. These splitters were tested exhaustively under deliberately harsh conditions and with markedly adverse materials characteristics using the Replicate Experiment approach,[3] resulting in the conclusion that both automated samplers can be fit-for-purpose, while also acknowledging the well-known fact that there is no such thing as a universal sampler matching all types of materials at all kinds of sampling conditions.

Representative Mass Reduction in the Laboratory

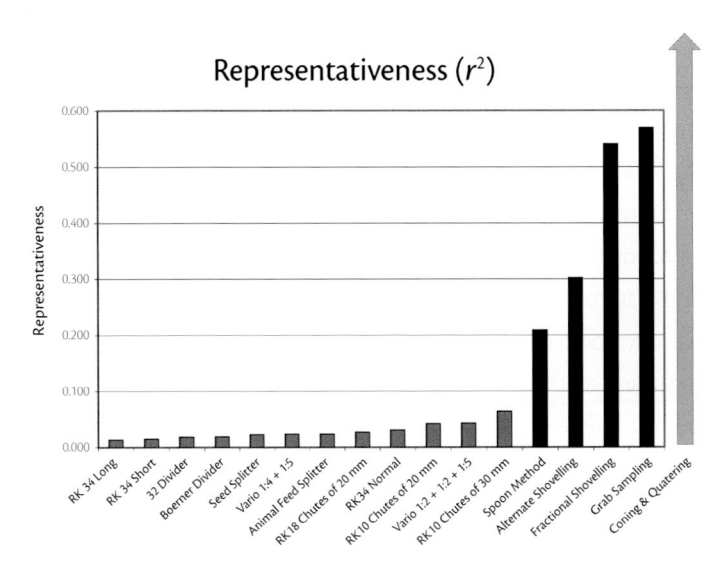

[Chart: Representativeness (r^2) bar chart ranking methods from lowest to highest: RK 34 Long, RK 34 Short, 32 Divider, Boerner Divider, Seed Splitter, Vario 1:4 + 1:5, Animal Feed Splitter, RK 18 Chutes of 20 mm, RK 34 Normal, RK 10 Chutes of 20 mm, Vario 1:2 + 1:2 + 1:5, RK 10 Chutes of 30 mm, Spoon Method, Alternate Shovelling, Fractional Shovelling, Grab Sampling, Coning & Quatering]

Figure 14.11. Ultimate representativity ranking of the 17 + 1 methods assessed in Reference 1. (The coning-and-quartering method was added after publication.[5]) Note that the performance of coning-and-quartering is dominantly a function of the degree of effective mixing preceding the "splitting" (using the world's lowest number of chutes, 4). This mixing is severely under-estimated, and is a critical success factor nearly always far less pervasive and complete than assumed. In rare occasions when these deficiencies have been carefully counteracted, coning-and-quartering can be compared with some shovelling methods (not that this helps anything, however).

14.6 Conclusions

So, mass reduction in the laboratory is anything but the easy matter of acquiring a piece of equipment that *claims* to be able to do representative sub-sampling, most often in the form of one or other form of sample *splitting*. Many do not—far from it: stringent performance documentation is always needed! Well there is one exception, which unfortunately cannot be applied to all types of material, but when this is the case, just order the Boerner divider—which can be used both for static dividing, or for dynamic in-line applications.

For all types of equipment that have passed muster in the representativity ranking, there is a rational set of rules that *must* be honoured in full in order for any alleged "splitter" to be representative. The most important of these have been introduced and illustrated above. An authoritative benchmark study allows anybody to perform a comprehensive audit of the state of a TOS application in their own analytical laboratory[1] and the magisterial textbook by Pitard goes into much more extensive details regarding all the principles to be mastered in order to eliminate the bias-generating Incorrect Sampling Errors (ISE) for a great many specific types of equipment.[2] A severe warning is sounded about coning-and-quartering,[5] a warning that is equally applicable at all scales.

14.7 References

1. L. Petersen, C. Dahl and K.H. Esbensen, "Representative mass reduction in sampling—a critical survey of techniques and hardware", in "Special Issue: 50 years of Pierre Gy's Theory of Sampling. Proceedings: First World Conference on Sampling and Blending (WCSB1)", Ed by K.H. Esbensen and P. Minkkinen, *Chemometr. Intell. Lab. Syst.* **74(1)**, 95–114 (2004). https://doi.org/10.1016/j.chemolab.2004.03.020, bit.ly/tos12-2
2. F.F. Pitard, *Theory of Sampling and Sampling Practice*, 3rd Edn. Chapman and Hall/CRC (2019). ISBN 9781138476486
3. DS 3077, DS 3077. *Representative Sampling—Horizontal Standard*. Danish Standards (2013). https://www.ds.dk, bit.ly/tos14-3
4. P. Wavrer, "An automatic linear proportional sampler based on the principles of the Theory of Sampling", *TOS Forum* **Issue 6**, 25–27 (2016). https://doi.org/10.1255/tosf.85, bit.ly/tos14-4

5. C. Wagner and K. Esbensen, "A critical assessment of the HGCA grain sampling guide", *TOS forum* **Issue 2,** 16–21 (2014). https://doi.org/10.1255/tosf.18, 👆 bit.ly/tos11-8
6. R.C. Steinhaus, J.J. Rust and M. Singh, "Does process control sampling always have to be a compromise?", in *Proceedings WCSB8*. AusIMM Publishing, pp. 197–202 (2017). ISBN 978-1-925100-56-3
7. M. Lischka, A. Holweg and K.H. Esbensen, "New online/at-line splitter designs for laboratory automation—feasibility results", in *Proceedings WCSB8*. AusIMM Publishing, pp. 159–166 (2017). ISBN 978-1-925100-56-3
8. R.C.A. Minnitt, K. Jakata and K.H. Esbensen, "The Grouping and Segregation Error in the rice experiment and at the assayer's bench", *Proc. WCSB9*, pp. 581–603 (2019). http://email.wcsb9.com/UploadFile/PROCEEDI_GS_OF_WCSB9-final20190520.pdf, 👆 bit.ly/tos14-8

15 Introduction to process sampling

Previous chapters presented introductions to the basic principles, methods and equipment for sampling of *stationary* **materials and lots**, as part of an initiation to the systematics of the Theory of Sampling (TOS). The next chapters will deal with **process sampling**, i.e. sampling from *moving streams of matter*. As will become clear, there is a great deal of redundancy regarding how to sample both stationary and moving lots, but it is the specific issues pertaining to *dynamic lots* that will be highlighted below.

> FSP: Fundamental Sampling Principle
> LDT: Lot Dimensionality Transformation
> FSE: Fundamental Sampling Error
> GSE: Group Segmentation Error
> TAE: Total Analytical Error
> MPE: Minimum Possible Error
> ISE: Incorrect Sampling Errors

15.1 Lot dimensionality: ease of practical sampling

The TOS has found it useful to classify lots into four geometrical categories. The strict scientific definitions are not necessary at the present introductory level, which will rather focus on lot dimensionality from the point of view of *sampling efficiency* (or sampling possibility, in difficult cases). A straightforward lot dimensionality classification on this basis can be seen in Figure 15.1.

From a practical point of view, sampling needs to be concerned with the ease with which one is able to extract increments from a *randomly chosen* location within the lot (or selected according to a systematic *sampling plan*). Thus, it is relatively easy to extract slices of any lot which has one dimension that *dominates*, i.e. is vastly longer (the extension dimension) than any of the other two dimensions (width, height). From this practical sampling point of view, the lot is effectively 1-D because

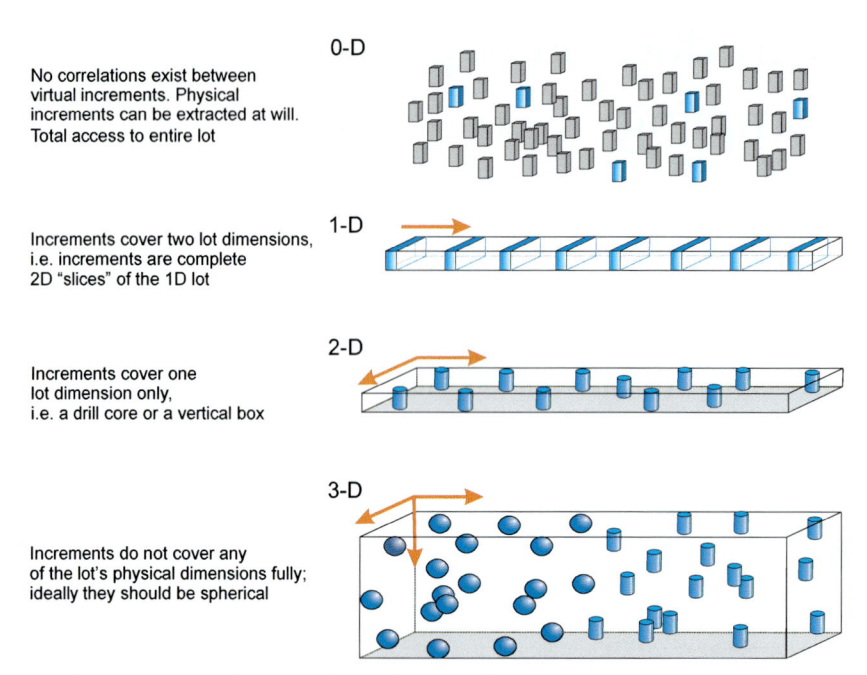

Figure 15.1. The TOS' practically oriented classification of lot dimensionality; the perhaps oddly named "0-dimensional lot" type is explained fully in the text. Blue signify extracted increments. Also see Figure 2.1 regarding alternative geometrical forms for increments for the 2-D and 3-D cases.

all material in a proper incremental slice "covers" the full width and height of the moving material stream completely. This is the reason for the TOS' classification of 1-dimensional lots, 1-D lots, or 1-D bodies. Observe that a 1-D lot can actually either be a *stationary*, significantly elongated body (stockpile etc.) or it can be a *dynamic* 1-D lot, i.e. a moving or flowing stream of matter. The material being transported by a conveyor belt is an archetypal dynamic 1-D lot; as is the moving matter confined to and ducted through a pipeline. It is critical that all 1-D lot sampling increments have the form of a complete "slice", see Figure 15.2. 1-D lots have a special status in the TOS, for the singular reason that this lot configuration makes for the easiest and most effective sampling conditions, no exceptions.

Introduction to Process Sampling

Figure 15.2. There should always be an element of randomness in a proper sampling procedure; here the locations *along* the extension dimension of a 1-D lot are selected in this fashion. All slices correspond to complete slices of the width–height dimensions of the lot. N.B. Most practical sampling of 1-D bodies takes place at regular intervals, or at fixed locations, along the extension dimension for reasons that will become clear in the following chapters.

It is equally easy to define a 2-D body. 2-dimensional lots are characterised by the fact that all increments will only "cover" one dimension completely. Very often 2-D lots are horizontal, with the remaining dimension vertical (think of a drill core penetrating a geological formation), or any other *layer*, though not necessarily in this flat orientation. The defining issue is that there are only two degrees of freedom regarding sampling, namely *where* in the X–Y plane is the vertical increment to be located—"where to sample in the X–Y plane?" The operative increments in sampling 2-D lots are either "cylindrical increments" (drill core sections) or box-like increments, compare Figures 15.1 and 2.1.

The key feature for the sampler, or for the sampling equipment, is that there is full access to the entire lot in the case of 1-D and 2-D lots. This is an important empowerment because it allows the demands of the Fundamental Sampling Principle (FSP) to be honoured with effective ease (barring *hidden* obstructions): All potential increments from a lot **must** be fully *accessible* for physical extraction if/when selected. Indeed, this feature is scale-invariant, one shall be able to sample **all** 1, 2-dimensional lots of any size under the guidance of the FSP.

Going on to 3-D lots leads to a perplexing revelation, however. It is very difficult to define a 3-D lot from the point of view of practical sampling, and logically the operative increment *form* should here be a *sphere*. But in our 3-D physical world, extracting spherical increments from solid bodies is pretty much impossible. Be this as it may, the TOS has many alternatives to offer for 3-D sampling, but this is mainly outside the present scope; however, see References 1–5.

What then is a "0-dimensional lot", the remaining top illustration in Figure 15.1? This is another of the TOS' empowering ways to focus on the underlying systematics of sampling. A 0-D lot is a lot that is "small" enough so that it is particularly easy, **always** and under **all** circumstances and with **all kinds** of appropriate equipment, to extract any size increment desired, *anywhere* (in practice increments of any form, so long as the increments are all *congruent*, i.e. of the exact same form and size). In other words, a 0-D lot is a small, particularly easy-to-sample lot. Obviously, there is a grading demarcation between a 0-D lot and a 3-D lot, since both **are** *de facto* 3-D lots in the physical manifestation when their obvious size differences are disregarded. But in practice the 0-D lot discrimination has been found immensely useful. It is the degree of *full accessibility* in sampling practice that is the key operative element in all these definitions.

Thus, with respect to sampling practice, lots come in groups [0-D, 3-D lots] vs [1-D, 2-D lots] of which the latter are of overwhelming importance—because these lot manifestations allow practical sampling no longer to be concerned with the size, volume or mass of the lot—*only* their heterogeneity matters. All 1-D, 2-D lots can be sampled appropriately, and this is a significant first step towards being in a position to be able to guarantee universal representativity in sampling.

15.2   Lot dimensionality transformation

The above is a most advantageous way of viewing the fourfold lot dimensionality array. Irrespective of whether a 1-D lot is stationary or moving, it is 100% guaranteed that the *entire lot* will be available for increment extraction. 1-D lots are **always** easy to sample, irrespective of their *original* configuration. They could have been 3-D or 2-D lots that were *transformed* into a 1-D configuration (for example under transportation). Universally then, from the largest lot sizes involved, e.g. a very big ship's cargo (100,000 tonnes) down to an elongated pile of powder in the laboratory—when present in a 1-D configuration, slicing the number of increments required, Q, constitutes the most effective sampling condition known. This scenario is the most desirable of all sampling options.

This finding has led to formulation of one of the six governing principles in the TOS, Lot Dimensionality Transformation (LDT). Wherever, whenever possible, it is a tremendous advantage to physically transform a lot (0-D, 2-D, 3-D) into the 1-D manifestation. Figure 15.3 illustrates this governing principle.

Even if there may be *some* work involved (N.B. occasionally this may be *a lot* of work in particularly adverse conditions) in

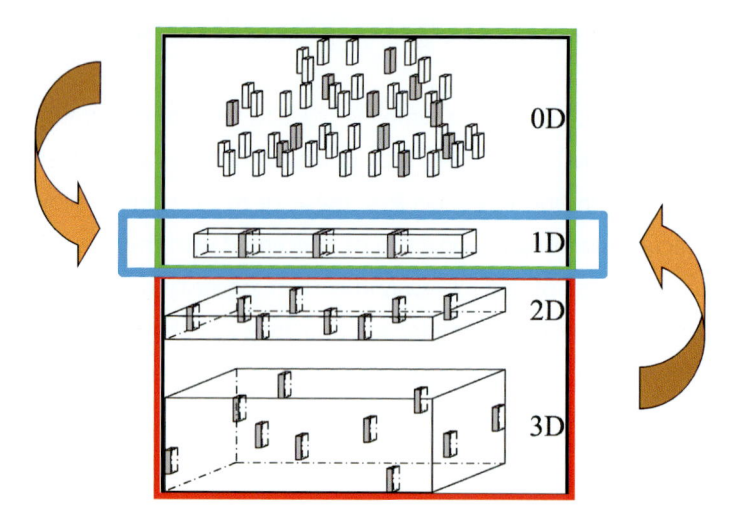

Figure 15.3. In practical sampling, The TOS has shown the highly desirable advantage of transforming 3-D, 2-D and sometimes even 0-D lots into a 1-D configuration.

moving, transporting (bit-by-bit) a lot, for example loading the complete lot content onto a conveyor belt, this is very often still a welcome resource expenditure because of the valuable sampling bonus now available. There is simply no comparison, because of the empowering ease with which the gamut of sampling errors can be eliminated or reduced with the 1-D configuration. The TOS' literature is full of examples, demonstrations and case histories on this key issue, e.g. References 1–5 and further literature cited therein. And this empowerment does not necessarily have to be based on automated equipment; Reference 6 is a tour de force regarding *barefoot sampling,* employing this idea with maximum efficiency under distinctly poor economic conditions.

Subsequent chapters will describe the power and diversity of *process sampling* in more detail; below follows a carefully selected first *overview*.

15.3 Process sampling

Process sampling concerns 1-D lots where there is a distinct spatial or temporal *order* between the prospective increments

Introduction to Process Sampling

along the defining extension dimension. These may appear either as an ordered series of discrete units (in time or space) or as a continuous moving/flowing material stream. All such elongated or moving material bodies are of course, strictly speaking, three-dimensional objects, but by transformation into 1-D objects their sampling turns out to be *identical* in principle as well as in practice. The movement involved is *relative*: either the matter streams, or flows, past the sampler/sampling equipment, or the sampler "walks up and down" along the extended dimension of the lot. From a sampling point of view, these two situations are identical and will therefore both be covered even through the terminology most often speaks of *process sampling*.

It is now time to focus on the nature of the lot material to be sampled. In process sampling, the 1-D lot can be classified in three broad categories:

- A moving, or stationary, 1-D lot of *particulate material*. Examples: conveyor belts transporting aggregate materials, interim or finished products, powders, heavy, dense slurries in ducts.
- A moving or stationary *fluid flow*. Examples: water in rivers or produced/manufactured fluids or diluted slurries ducted in pipelines.
- A moving or stationary stream made up by a series of *discrete units*. Examples: railroad cars, truck loads or bags, drums, packages... from a production or a manufacturing line.

Besides the distributional and constitutional heterogeneity (introduced in earlier chapters), there are further aspects that need to be considered to characterise the heterogeneity of 1-dimensional lots. This especially involves understanding the nature of the non-random heterogeneity fluctuations *along* the elongated lot. Interest is no longer so much about the heterogeneity *within* increments, because a full slice will be extracted and its heterogeneity is, therefore, now only a matter for the subsequent mass-reduction step(s), an issue easily managed under the TOS. The lot heterogeneity of interest is now *only* the heterogeneity *along* the entire length of the 1-D lot, i.e. within the "linearised" moving *volume* of the original lot.

Often "slicing" in such a case amounts to nothing more than appropriate *selection* of already prepared units viewed as a basis for a time series of analytical results, i.e. units may be manufactured units, e.g. bags, cans, containers or one can jump up in scale, truck loads, train car loads etc. But the 1-D lot will also often manifest itself as a more or less continuous body (1-, 2-, 3-phase continuum) along the length dimension, in which case the sampler/sampling equipment must forcefully "cut" a slice across the stream to produce the extracted units (increments). The location of **where**, and **how**, to cut the stream of matter is of critical importance in process sampling. This active element even influences the terminology of sampling—a cross-stream sampler [a traversing cutter (see front cover of this book) or a hammer sampler are examples] is also known by the generic term, a "cutter".

Critical distinction
A cross-stream sampler (cross-stream cutter) intersects a falling stream of matter. A cross-belt sampler (or a cross-flow sampler, ducted flow) tries to do this feat by crossing an active conveyor belt (or sampling a slice of a ducted flow). The latter types are facing orders of magnitude of more complex hindrances before getting close to the possibility of being correct. Cross-stream sampling is the by far most advantageous option! See References 1–6 for in-depth descriptions.

The heterogeneity contribution *along* a 1-D body is composed of four parts, with respect to the specific form of the compositional variation:

1) A random, discontinuous, short range fluctuation part. This term describes the constitutional heterogeneity *within* an increment. If the movement of the 1-D body could be stopped, this heterogeneity gives rise to the Correct Sampling Errors, FSE and GSE *only* (see later chapters for details).

2) A non-random, continuous, long-range fluctuation part that describes a *trend* in the process/lot (between units) over time/distance (if such is present).

3) A non-random, continuous, cyclic part, describing cyclic or *periodic* behaviour of the 1-D process/lot (if such is present).

4) A random fluctuation part, taking into account all error effects stemming from weighing, sample processing and analysis. This can be viewed as the *extended* TAE. Sometimes it is desired to keep the strict analytical errors isolated, as TAE proper. Either way, no confusion need arise and various cases will be illustrated in the following chapters.

1-D lots may display all four types of heterogeneity along the transportation direction, but trends and/or periodic behaviour may, or may not, be a characteristic, depending on the nature or origin of the lot to be sampled in transport. Many manufacturing, and some material handling processes will result in a more-or-less developed trend or a periodic heterogeneity characteristic, which usually are *tractable* (understandable) in their technological and industrial contexts. Many natural processes will have more complex, overlapping characteristics with a wide-ranging variation in the degree of their individual manifestations.

15.4 1-D lot heterogeneity

Characterisation of the heterogeneity of a 1-D lot must include information on the *chronological order* of the units extracted and their *in-between* correlations. Upon reflection, it is clear that it will be of interest to be able to characterise the intrinsic heterogeneity of the 1-D lot at *all scales* from the increment dimensions upwards. There can be no *resolution* of the 1-D heterogeneity characterisation *smaller* than the physical dimension of the increment width in the extension direction (which is defined as the *lag*, see below), but all larger scales are of principal interest, all the way up to half the length of the entire 1-D body (corresponding to the, in practice, unlikely case in which the lot was sampled at only three locations at the first, the central and the last position, respectively). It is of great interest to be able to express the 1-D lot heterogeneity at all these scales *simultaneously*. This may appear as a complex task, but the TOS has developed an amazing facility for exactly this purpose—the *variogram*.

15.5 Variographic analysis: a first brief

In order to characterise *autocorrelation* between units of a process/lot, the variogram is very powerful. It allows the variation observed between extracted increments to be understood as a *function of the distance between them*, in time or space. The smallest equidistance between increments to be extracted is called the "unit lag". There will always be a minimum lag distance which is to be determined by the sampler when setting up the basis for a variographic characterisation (the variographic data analyst). In many cases (but certainly not all) valuable information already exists, from general process experience or from an earlier attempt to map the characteristics of the lot in question, which will allow a first try minimum unit lag to be fixed with particularly high relevance.

A variogram will yield information in the form of the so-called "nugget effect", the "sill" and the "range", which are introduced below.

A variogram is based on the analytical results from a *series* of extracted increments, which subsequently are all mass-reduced and analysed in an exactly identical TOS-correct manner, so as to suppress as much as possible sub-sampling, mass-reduction and analytical error effects from influencing optimal comparison of the between-increment relationships. All extracted increments are in a sense first treated as individual "samples" in the variographic context (but this initial status as *grab samples* is not a cause for worry, as will become clear).

A *variogram* can be calculated based on a series of analytical results from a *necessary and sufficient* number of increments, spanning the entire process interval of interest. An example could

Introduction to Process Sampling

be a production process over a 24-h period, for example sampled every 20 min (lag = 1) to characterise the variation over three 8 h shifts. This would yield a total of 72 analytical results. The issue of how to fix an appropriate number of data from which to calculate variograms is also addressed in later chapters. Here, it is sufficient to state a beginner's rule-of-thumb: preferentially 60–100 data points (analytical results) must always be collected! This being the rule, it should be said, however, that *sometimes* shorter series *can* also be investigated—or *sometimes* the variographic experiment may employ a much higher number, like daily or seasonal variation, for periods up to months, an entire year or even longer. In general, the variogram is supposed to characterise a salient "process interval of interest". This is, of course, an interval that is intimately related to the *specific* process in question, but the common feature is that the process must be covered with *at least* 60 increments.

The fundamental operative unit used in variogram calculations is the lag parameter, j, describing the systematic distance between two extracted units. There is a smart way to cover all inter-unit distances of interest in a fully comparable manner: *larger lag lengths* (than 1) are expressed as a dimension-less, relative lag derived by expressing these as a series of *multipla* of the basic minimum lag = 1 unit. Thus, the lag runs in the interval [1,2,3,4,5, … 36] in the example above, or in the interval [1,2,3,4,5, … $N_U/2$] in the general case in which the total number of unit extracted is N_U. This simple re-expression makes variographic interpretation and comparison of **all** processes possible with the greatest ease.

15.6 Interpretation of variograms

The *interpretation* of the physical meaning of variograms is the first important result of a variographic analysis. The variogram *level* and *form* provide specific information on the process variation captured (or, the systematics of the 1-D heterogeneity in the case of a stationary 1-D body). Normally, four primary types of variograms are encountered only (based on the TOS' ~60 years of wide experience):

- The *increasing variogram* (normal variogram shape); this variogram has a *range* (see below).
- The *flat variogram* (no autocorrelation along the defining dimension); this variogram has **no** range (see below).
- The *periodic variogram* (manifested as a *superposition* on either of the first two types).

These variograms are all outlined schematically in Figure 15.4. When the variogram type has been identified, information on

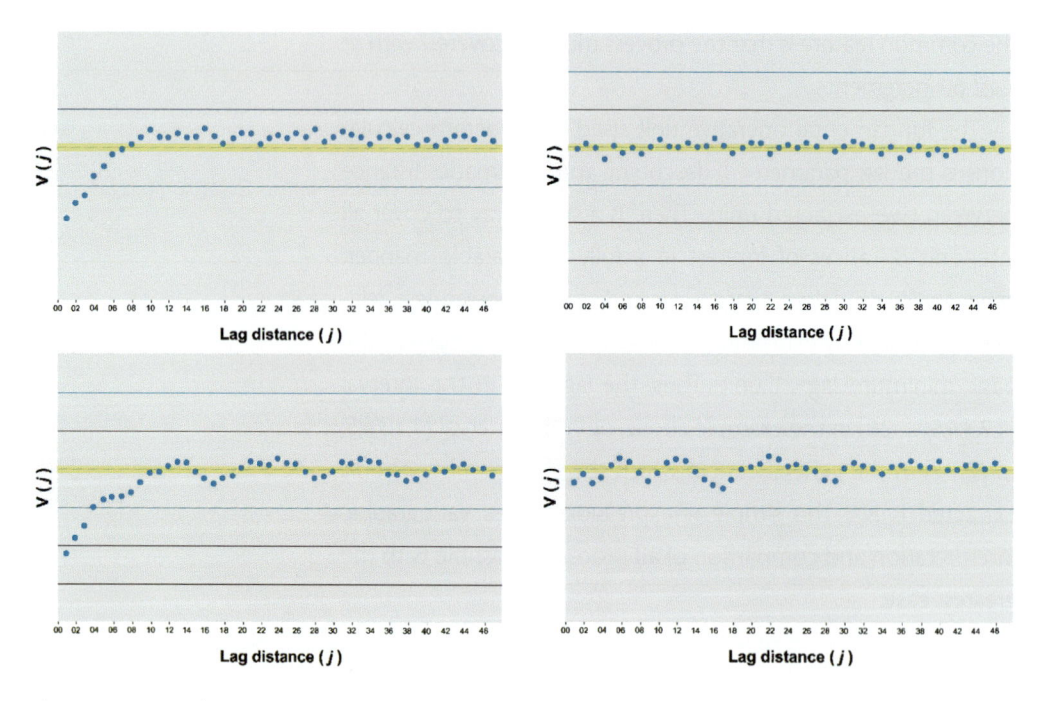

Figure 15.4. Four basic variogram types. Top panels show the increasing and the flat variogram. Bottom panels show a periodic overprint on both these types. Strictly speaking, a flat variogram does have an increasing segment, only this is so small that the range is smaller than the unit lag chosen. This would be revealed, for example, by conducting another variogram with a much smaller lag unit.

further optimisation of routine 1-D sampling can be derived (and there are many other types of information that can be gained from variograms). The increasing variogram (Figure 15.4, top-left variogram) will be used as an example for the present introduction.

Variograms are not defined for lag $j = 0$, as this would correspond to extracting the exact same increment twice. However, even though this is not physically possible, it is still highly valuable to acquire information as to the *expected variation* corresponding to this situation (sometimes called the zero lag variance). The TOS identifies this variation as the "nugget effect" (also termed the "minimum possible error", MPE). Normally, the first five points of the variogram are extrapolated backwards to *intercept* the ordinate axis to provide an estimate of the magnitude of the nugget effect (but slightly more complex *model* curve-fitting alternatives also exist, some of which even force the variogram to go through zero for the virtual case of lag = 0; these are the preferential choices within *geostatistics*). The TOS way of back-extrapolation furthers the easiest *estimate* of the Y-axis intercept which still carries a wealth of relevant information.

There is a reason for the name "MPE", because the nugget effect/MPE in fact includes uncertainty contributions from all error types influencing sampling-and-analysis measurement systems if not sufficiently TOS-optimised. Thus, MPE includes both significant Correct Sampling Error effects (FSE, GSE), as well as a full complement, or the remaining residual complement, of Incorrect Sampling Errors (ISE) in addition to TAE, all contributing to an elevated, *unnecessarily inflated* nugget effect if not addressed properly. MPE is, therefore, an appropriate measure of the absolute minimum error that can be expected in practice after having employed the full complement of sampling error elimination and reduction measures available in the TOS—while sampling and measuring with the fastest frequency possible corresponding to

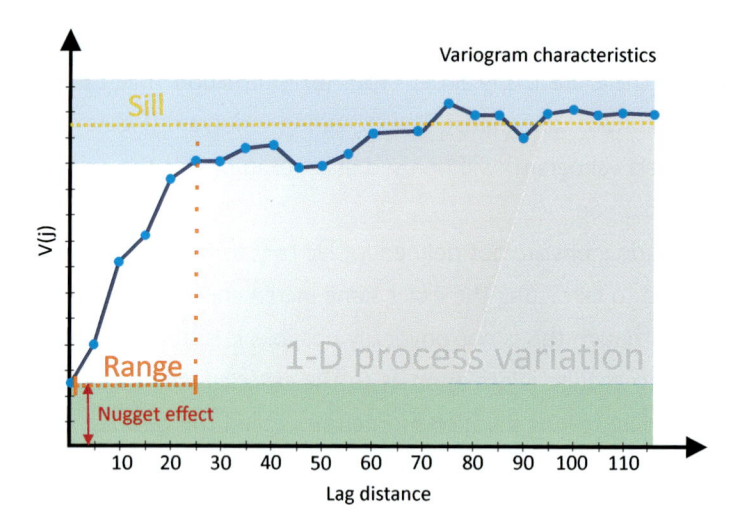

Figure 15.5. Generic increasing variogram, schematically defining the *nugget effect*, the *sill* and the *range*. These three parameters suffice to characterise all types of variograms, *cf.* Figure 15.4

the minimum lag = 1. While not necessarily an optimal solution for all practical technological and industrial scenarios ("overkill")—typically because of a desire to keep the costs of sampling systems as low as possible—it is of vital importance to know this datum when interpreting the behaviour of real-world measurement systems. Several examples will be given below.

Figure 15.5 shows a *generic* increasing variogram, delineating the three basic variogram parameters, *nugget effect*, *range* and *sill*, which is all that is needed to characterise **any** variogram.

When the increasing variogram becomes more or less flat after a certain multiplum of unit lags (X-axis), the "sill" of the variogram has been reached. The sill provides information on the observable *maximum* sampling variation. N.B. the left-ward "dip" of a smoothed version of an increasing variogram signifies an *increase* of between-unit autocorrelation as the lag becomes smaller and smaller (classical definition of time-series autocorrelation). The "range" of the variogram is found as the lag beyond which there is no longer any discernable autocorrelation between increments, i.e. when the variogram has become flat. TOS process sampling is

Introduction to Process Sampling

Figure 15.6. Manual increment extraction from a conveyor belt defining a dynamic 1-D lot. The scoop used to extract increments is less than half the width of the conveyor belt, imparting significant, uncontrollable Incorrect Sampling Error effects to the process sampling. This results in an (unnecessarily) *inflated* nugget effect, which is one of the means by which variographic process characterisation can also be used for total sampling-plus-analysis system evaluation, see References 1–5 and further references therein. Photo credit: KHE Consulting.

mostly interested in what takes place below the range, i.e. what characterises pairs of increments with a smaller between-unit distance than the range. This will be developed in later chapters.

If a significant periodicity is observed, the amplitude of which will be dampened at higher and higher lags (Figure 15.4, lower variograms), the sampling frequency must **never** be similar to this period, since this risks introducing an additional error, an in-phase error, in which the unit extraction unluckily is in phase with the periodic process variation. In these cases, the specific sampling mode (random sampling or stratified random sampling) becomes critically important.

A complete introduction to variographic characterisation and process sampling is no small matter, and the present initiation must be complemented by a more substantial learning effort. To whet the reader's appetite, see Figure 15.6.

For the impatient reader, recent comprehensive introductions to variographics can be found in References 1–4, also containing a selection of further background literature. For a complete introduction, see also the TOS' most recent high-level textbook.[5]

15.7 References

1. DS 3077 (2013). *Representative Sampling—Horizontal Standard*. Danish Standards (2013). http://www.ds.dk, bit.ly/tos12-3
2. L. Petersen and K.H. Esbensen, "Representative process sampling for reliable data analysis—a tutorial", *J. Chemometr.* **19**, 625–647 (2005). https://doi.org/10.1002/cem.968, bit.ly/tos15-2
3. K.H. Esbensen and P. Mortensen, "Process sampling (Theory of Sampling, TOS) – the missing link in process analytical technology (PAT)", in *Process Analytical Technology*, 2nd Edn, Ed by K.A. Bakeev. Wiley, pp. 37–80 (2010). https://doi.org/10.1002/9780470689592.ch3, bit.ly/tos6-6
4. R. Minnitt and K. Esbensen, "Pierre Gy's development of the Theory of Sampling: a retrospective summary with a didactic tutorial on quantitative sampling of one-dimensional lots", *TOS forum* **Issue 7**, 7–19 (2017). https://doi.org/10.1255/tosf.96, bit.ly/tos6-13
5. F.F. Pitard, *Theory of Sampling and Sampling Practice*, 3rd Edn. Chapman and Hall/CRC (2019). ISBN 9781138476486
6. *TOS Forum* **Issue 9** (2020). https://www.impopen.com/tosf-toc/19_9, bit.ly/tos15-6

16 Process sampling: the importance of correct increment extraction

After the previous chapter's introduction to the *why*, the *how* and the *technicalities* involved in process sampling and variographic analysis, it is time for a bonanza of applications and case histories covering as broad a practical scope as possible. In this chapter, we introduce the critical *prerequisites* for the *variographic experiment*, by focusing on the importance of TOS-correct increment extraction for proper variographics. This issue cannot be over-emphasised.

A
DH: distributional heterogeneity
ISE: Incorrect Sampling Errors
IDE: Incorrect Delineation Error
IEE: Incorrect Extraction Error
IPE: Increment Preparation Error
IWE: Increment Weighing Error

16.1 Moving, or static, 1-D lots: increment cutting must be TOS-correct

Figures 16.1 and 16.2 illustrate how focus is on the extension dimension in process sampling (aka one-dimensional, 1-D sampling), as long as each increment complies with the TOS' stringent demand for a *complete slice* of the two width–height dimensions. By securing increments of this geometric configuration, there is only the extension dimension heterogeneity left, i.e. the longitudinal in-between increment spatial heterogeneity (DH = distributional heterogeneity). All variographic characterisation is aimed at describing and managing $DH_{process}$.

If this demand is not observed, see Figures 16.3 and 16.4, it is clear how there will be a fundamental compositional *imbalance* from one increment to the next, which should not be used for the purpose of characterising the 1-dimensional DH.

Figure 16.1. Dynamic (moving) 1-D lots, from left to right: a conveyor belt transporting coal to a power plant, a pipeline, a series of produced goods). Photo credit: Hans Møller, with permission, and KHE Consulting.

Figure 16.2. Stationary 1-D lots produced in a laboratory-scale study of internal distributional heterogeneity (DH) as a function of the specific stacking process. Observe how it is critical to acquire a complete cross-sectional slice of the lot due to the hidden heterogeneity, which can be very different from the particular surface manifestations. Photo credit: KHEC teaching collection.

Planar–parallel or curvy-parallel cross-sections of a moving stream are the **only** TOS-correct delineations of physical process sampling increments, Figure 16.5 (cases "A" and "C") and Figure 16.6 top panel, eliminating a potential incorrect delineation error [one of the four potential incorrect sampling errors, ISE].

Incorrect increment delineation, and extraction, will give rise to an inflated nugget effect in variographic process sampling characterisation. Non-compliance with these basics will give rise

Process Sampling: the Importance of Correct Increment Extraction

Figure 16.3. Examples of irregular, unacceptable, partial cross-section slices compared to a correct full slice of the moving stream; the first two (from right to left in the figure) do not comply with the definition of correct process increments. Observe also the marked asymmetrical load on the conveyor belt, which would wreak havoc with either of the partial cross-sections. The insert shows when this adverse issue is taken to its extreme, with no observance of the need of securing a balanced cross-section of the moving stream. Indeed, here the objective seems to be "get a full bucket with the least hassle", which most certainly does not make for representativity—it is grab sampling plain and simple. Photo credit: KHEC teaching collection.

Figure 16.4. Evidence of a very haphazard, irregular "sampling", in effect *grab sampling* from one side of a conveyor belt only, cf. Figure 16.3. Also shown is a trace of what would have been a correct cross-stream increment. Photo credit: KHEC teaching collection.

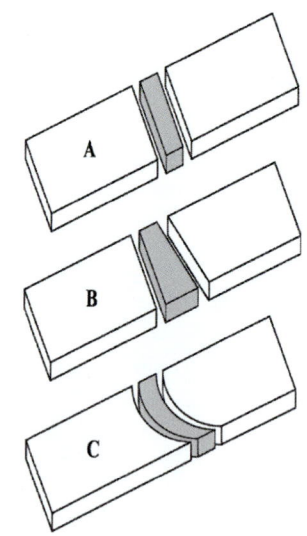

Figure 16.5. Illustration of the "stopped belt" situation in which it is possible to extract a perfect, TOS-correct cross-stream slice (central photo). The geometric delineation may be as in "A" or "C", but not like "B", which depicts an unbalanced cross-cut. Example taken from a calibration test of an on-line sampler at a coal power plant, to be validated against the representative samples extracted as illustrated in the centre. Photo credit: Hans Møller; with permission.

Figure 16.6. The "complete stream slice" dictum: only a complete cross-section of the moving stream of matter will satisfy the TOS' principles for correct increment extraction. This illustration can be thought of both as looking down on the top surface of a conveyor belt, or as a longitudinal cross-section of a pipeline. Note carefully the "oblique" trace (right-most slice) which is the practical realisation of all cross-cutter samplers when sampling from a moving stream. Illustration credit: KHEC teaching collection.

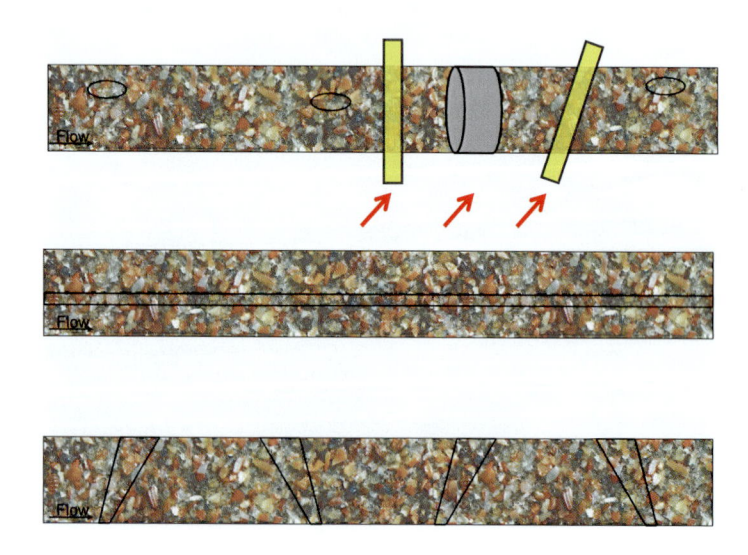

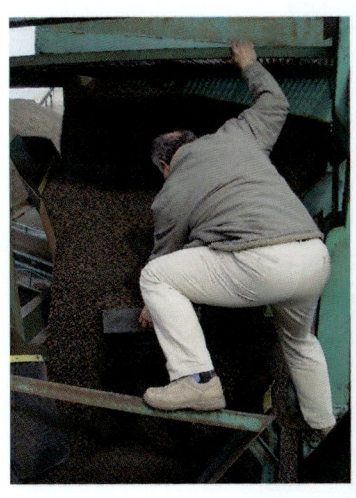

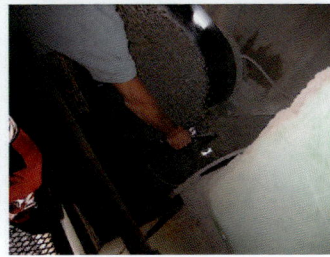

Figure 16.7. It all starts here! Horrific examples of extremely unbalanced, IDE/IEE-ridden attempts of manual increment process sampling—always doomed to fail. Manual process sampling is pretty much always a fatal surrender to practical and economic complacency, at the cost of a massive sampling bias, which in turn is bound to incur completely uncontrollable hidden costs. Photo credit: KHEC teaching collection.

to incorrect sampling errors (IDE; IEE: Incorrect Delineation Error; Incorrect Extraction Error) which are **unnecessary,** and which **can** in fact be eliminated from the sampling process.

Manual increment extraction is nearly always a bad idea, and can never be TOS-correct in practice; see horrific examples in Figure 16.7. Whenever all Incorrect Sampling Errors (IDE, IEE, IPE, IWE) have not been eliminated, the sampling process will invariably be fraught with a fatal, inconstant *sampling bias*, which can be never be corrected for; see further below.

There is no excuse for not getting the fundamental increment sampling right—and right from the start. Figure 16.8 shows three examples that are all completely TOS-correct, which is the first condition for process sampling representativity.

Only representative increment sampling processes are of interest in science, technology and industry. All examples shown in Figure 16.8 allow proper variographic process characterisation. Any violation of these simple requirements affects a given

Figure 16.8. Examples of TOS-correct increment delineation/extraction. Left: a "fish stair case" transporter is functioning as a correct increment "cross stream" cutter (even if the "unit particles" in this case are more than unusually large). Centre: the bottom outlet opening of a grain off-loading hopper also functions so as to delineate increments without IDE/IEE. Right: a role model "cross-stream" cutter at work at the terminal end of a conveyor belt. All increment cutting operations shown here are representative. Credit: KHEC teaching collection.

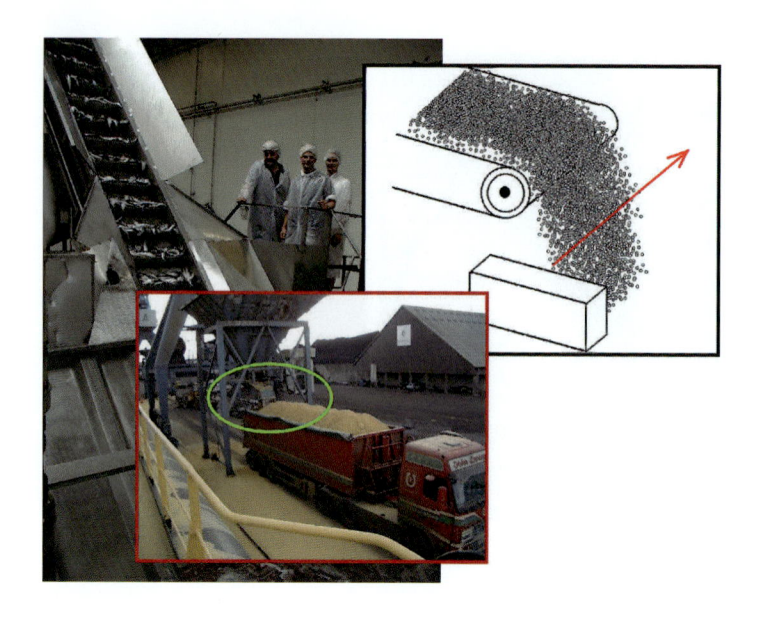

sampling process and will lead to an unnecessary inflated nugget effect, see further below.

Observe above how 1-D lots of both types, static and dynamic (moving) must live up to the *same* demands concerning the fundamental increment cutting requirements, as illustrated in Figures 16.5 and 16.6.

16.2 "Sooner or later"...

Sooner or later, however, the scene will be ready to perform proper process sampling—enter *variographics*. The variogram was introduced in the previous chapter in some detail, so most of what is lacking is simply a basic understanding of what the variogram portrays with respect to the process, and how this comes about. A very small matter of one mathematical equation is all it takes: the professional sampler *must* understand the meanings and implication of the variographic master equation, which will be revealed in the next chapter.

For readers who have been inspired to know more about variographics, and who cannot wait, there is salvation in the two standard references,[1,2] as well as the new professional introduction published recently by Minnitt & Esbensen.[3] This latter also takes you through the full mathematical intricacies; well suited as a follow-up to the present introductory chapters.

This chapter has presented the critical issue of correct increment delineation and extraction in great detail (eliminating the otherwise fatal Incorrect Sampling Errors contributing to a sampling bias)—for a good reason. Full attention to these issues is essential before embarking on the powerful variographic process characterisation. The next two chapters are filled with practical case histories in which both benefits and unnecessary adverse effects will be revealed.

16.3 References

1. DS 3077. *Representative Sampling—Horizontal Standard*. Danish Standards (2013). http://www.ds.dk, bit.ly/tos16-1
2. K.H. Esbensen and C. Wagner, "Theory of Sampling (TOS) versus Measurement Uncertainty (MU)—a call for integration", *Trends Anal. Chem. (TrAC)* **57**, 93–106 (2014). https://doi.org/10.1016/j.trac.2014.02.007, bit.ly/tos16-2
3. R. Minnitt and K. Esbensen, "Pierre Gy's development of the Theory of Sampling: a retrospective summary with a didactic tutorial on quantitative sampling of one-dimensional lots", *TOS forum* **Issue 7**, 7–19 (2017). https://doi.org/10.1255/tosf.96, bit.ly/tos16-3

17 The variographic experiment

Pierre Gy, the founder of the Theory of Sampling (TOS), pioneered applications of *variographics* to understanding meso- and large-scale variability systematics in industrial processes, process plants and for process control purposes from as early as the 1950s, and devoted a major part of the TOS' development to this subject. The *variogram* allows one to differentiate *sources of variability* and provides valuable insight due to correlations between successive samples from a moving streams. Neglect or poor understanding of the data analytical capabilities of the variogram means that it has not been widely applied in process control, except in particular industry sectors which embraced the TOS early on and which are now the historical application sectors: mining, cement, coal/power generation and certain parts of other commodity processing industries. This was because of the overwhelming consequences of making uncertain decisions when treating large volumes and tonnages—the consequences of wrong decisions were simply too great. Failure to address stream heterogeneity in a proper context means that conventional statistics and Statistical Process Control (SPC) too often fail to identify and distinguish the underlying *sources of variability* in a process stream. For each type of heterogeneity, there is a matching component of process variability. Although the variographic approach is extremely powerful in terms of the insights one is able to gain in regard to process performance and management, examples of the application of this method have been suspiciously absent in the mainstream process technology literature.

17.1 The variogram

Any process stream that is to be sampled should always first be subjected to a "variographic experiment", the purpose of which a.o. is to *tune in* an optimised sampling frequency based on the increment size selected. The variographic experiment will also help to estimate an optimal number of increments to be aggregated as composite samples. In a variographic experiment it is the responsibility of the sampler to come up with the best possible initial suggestion for the *size* of the increments to be used; obviously, previous experience and knowledge regarding the specific process at hand are of great value. But simple trial-and-error experiments are also perfectly acceptable, fast ways into this complex matter—which is actually complex **only** if one was deliberately searching for a theoretical solution.

In order to characterise a process stream, it is necessary to extract a certain number of increments, N_U, to have these analysed in the laboratory and to conduct calculations based on the variographic master equation, Figures 17.1 and 17.2. (For this discussion we can assume that the analytical laboratory knows its responsibility well w.r.t. proper, representative mass-reduction.) The total number of analytical results (stemming from the N_U increments) *should* preferentially be between 60 and 100—but it may well be larger (this is actually not such a harsh demand, when it is factored in that most of the variographic characterisations used extensively in science, technology and industry are usually realised based upon *automated sampling*). In general, N_U *should not* be smaller than 60, although experienced operators occasionally cite the canonical number 42 as an absolute minimum (this is not recommended, however, without *considerable* experience).

The sampling frequency used in the variographic experiment is either set by the process situation at hand (based on existing, proven knowledge), or it may be calculated as the total process

The variographic experiment

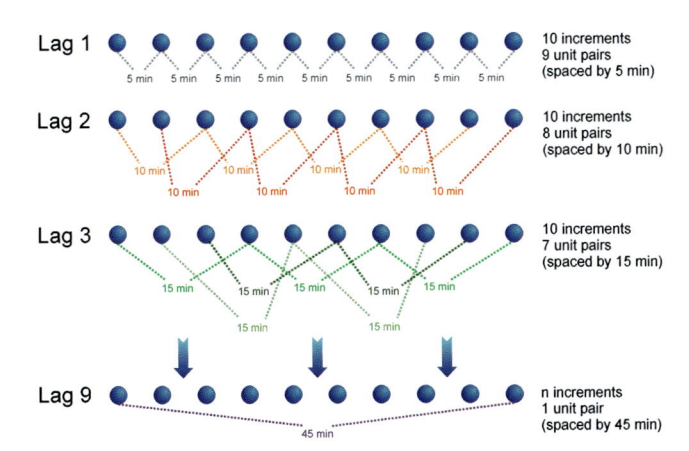

Figure 17.1. Based on a relevant problem-dependent sampling frequency, usually 60–100 increments need to be extracted, each extracted in a completely TOS-correct fashion. In the particular example shown here the sampling frequency is 5 min. Note how different multiple lags (distance between increments pairs) can be derived from the same basic sequence of N_u serial analytical results. No additional sampling is necessary for multi-scale variographic data analysis, see text.

interval under investigation divided by 60 (or 100). Often there are special circumstances that determine this choice, for example in the case where the variographic experiment is aimed at investigating a current situation, which has an already set sampling frequency. This may be defensible, or it may not—a matter that will be revealed by proper interpretation of the variogram results (several examples to follow in subsequent chapters).

There may be many different objectives behind a variographic characterisation, but all involve deciding upon the most relevant sampling frequency from which to gain a maximum of insight; more on these initiating issues after a first familiarity with the variographic experiment has been gained.

Thus, there is a minimum *resolution limit* associated with every variographic experiment; there can be no information gained at a scale *less* than the experimental sampling rate, the unit lag = 1, in the example in Figure 17.1.

Figure 17.2. Variogram master equation, expressed both in terms of conventional analytical results, a_m, or alternatively, in terms of the corresponding heterogeneity contributions h_n (the latter was defined in an earlier chapter, but repeated here for easy reference).

$$V(j) = \frac{1}{2(N_u - j)} \sum_m (h_{m+j} - h_m)^2$$

$$V(j) = \frac{1}{2(N_u - j)a_L^2} \sum_m (a_{m+j} - a_m)^2$$

$V(j)$ = Variogram function [relative (h_m) or absolute (a_m)]
h_m = Increment heterogeneity contribution

$$h_n = \frac{a_n - a_L}{a_L} \times \frac{M_n}{\overline{M}_n}$$

a_n : increment concentration
a_L : lot grade (process average)
M_n : increment mass

The distance between two data points is called the lag, j. The minimum distance between any two data points is termed Θ_{min}. Any distance between pairs of data points, j, is referred to this base Θ_{min}, and will therefore always be a *multiplum* of Θ_{min} [j = 1, 2, 3, ..., $N_U/2$]. This allows use of a general lag parameter, j, which is *independent* of the particular measurement unit used. This is a most welcome feature, allowing *comparison* between variograms of different processes, different process types, different materials etc. As shall be shown this makes comparative variographic analysis indispensable in process technology and process sampling, and this is the reason that professionals always prefer the relative variogram based on heterogeneity contributions.

It is often recommended to *over-sample* for the purpose of a first-go variographic experiment. Thus, in Figure 17.1 the current sampling frequency was actually ~15 min, but it was decided to over-sample by a factor of ×3, because there was an indication in the historical records that the current period was possibly too high.

The primary job for variographic characterisation, Figures 17.3 and 17.4, is to express the variability systematics of the set of N_U analytical results. Remember that due diligence (TOS correctness, i.e. unbiased sampling) must always be observed regarding

The variographic experiment

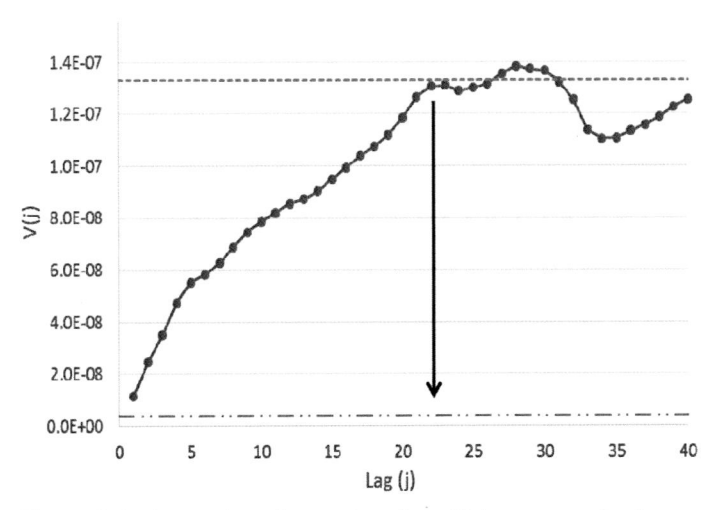

Figure 17.3. A generic *variogram* based on 80 increments (each representing a 2-minute process interval). This is a real-world variogram. The lag axis of a variogram will always be of length $N_U/2$. This variogram shows a very small nugget effect (dotted and dashed horizontal line); the sill is marked with a dashed line. The range is of the order of 22–23 lags, i.e. ~45 min, a feature being interpretable in the context of the process behind the variogram. Note that it is also known from earlier studies of longer duration than the present, that the sill indeed corresponds to the level shown here. This is an example of bringing in full domain-specific knowledge and experience in interpreting a variogram. See Figure 15.5 for explanation of the technical terms used here (nugget effect, range and sill).

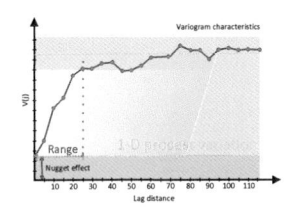

Figure 15.5. Generic increasing variogram, schematically defining the *nugget effect*, the *sill* and the *range*.

extraction of increments (see previous chapter). Indeed, the same adherence to the TOS' principles must, of course, also be observed for all the subsequent sub-sampling and sample preparation steps in the laboratory. On this basis, the only operative variability left is that *between* analytical results in the extension dimension (the process dimension). Thus, the variogram is a powerful characterisation of the 1-D *longitudinal heterogeneity* of the process under consideration because all *transverse* heterogeneity w.r.t. the process direction has been *covered*, i.e. *incorporated* in each increment extracted. N.B. Although in a variographic experiment it is a set of *increments* that are extracted,

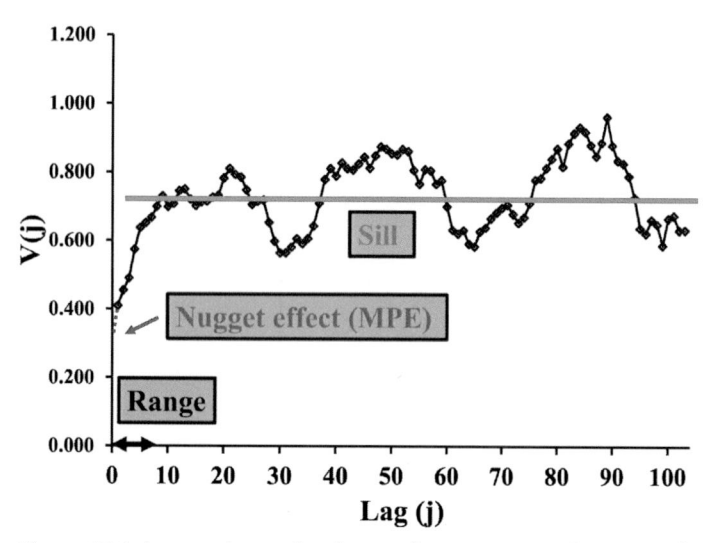

Figure 17.4. An experimental variogram from a process of great significance in technology and industry, *mixing*. The process data behind this variogram corresponds to control samples extracted at the mixer outlet. Note that the original data series is larger than 200 increments. The observable degree of periodicity in the variogram shows with clarity that the mixing process has **not** gone to completion. The performance of this mixer would not be accepted based on the evidence from this variogram. This illustration also shows typical practical manifestations of the basic variogram parameters, sill, range and nugget effect.

they are at first treated as *samples* in their individual right. The result of a variographic experiment *may* subsequently result in a certain number of increments having to be aggregated to form composite samples.

The variogram principle is to calculate the *average of the squared differences between all pairs of data points with in-between spacing equal to the lag, j,* as j runs the interval of interest. Thus this fundamental calculation is *repeated* for all j lags, i.e. [j = 1, 2, 3, ..., $N_U/2$].

Figure 17.1 shows the spatial disposition of all possible pairs of data as a function of the increasing lag [j = 1, 2, 3, ..., $N_U/2$].

The variogram master equation returns a singular value, the variance V, for each lag in question, V(j), i.e. there is calculated a single variance measure *corresponding* to each lag. The variographic function thus characterises the full set of N_U data by a lag-variance duplet "one scale at a time" [j = 1, 2, 3, ..., $N_U/2$]. Plotting V(j) [Y-axis] as a function of the lag j [X-axis] then produces the *variogram*, Figures 15.4, 15.5, 17.3 and 17.4.

For newcomers to variographics there may appear to be an apparent ambiguity regarding whether to express the variogram based on absolute concentration values, or recalculated as heterogeneity contributions. Indeed, within some industry sectors certain robust traditions can sometimes be met with, but these are just that: *idées fixes*. Better to use the relative variogram. Figure 17.2 shows both options, termed the absolute vs the relative variogram, respectively. This is a matter of no consequence, however, as the *shape* of the alternative variograms will be *similar*, with only the unit of the original measurements (and thus the unit on the Y-axis variances) differing. Interpretation of both types of variograms will be *identical*. The advantage of using the relative variogram is significant, as it allows direct comparison of all variograms *inter alia*, including the levels and magnitudes of *ranges*, *sills* and *nugget effects*, an important point to be further explored.

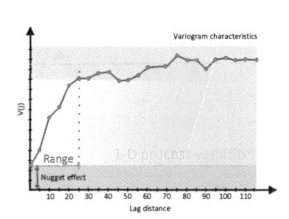

Figure 15.5. Generic increasing variogram, schematically defining the *nugget effect*, the *sill* and the *range*.

Based on the present and the preceding two chapters, the reader is now ready for a selection of typical and illuminating real-world examples and case histories from which to learn some of the powerful capabilities of *variograhics*.

Figure 17.4 is a real world variogram from a technological process. The sill is often considered as a kind of *ceiling* for the total variability across the full lag interval—*technically*, however, the sill is defined as the average variance for all lags. In well-prepared variograms with a sufficient number of increments in which the *range* usually only constitutes a small number of lags, the average variance (the sill level) will be lowered by that part of the

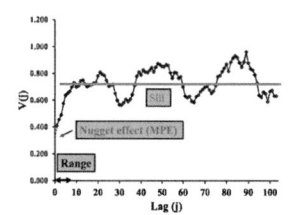

Figure 17.4. An experimental variogram...

variogram that is within the *range*. Note that the ceiling will not cap the variability from above, but will asymptotically envelope the data series from below.

As soon as the lag distance goes beyond the range, the particular variogram in Figure 17.4 shows a tell-tale periodic pattern with a period of ~30 lags superposed on a flat sill. The process being characterised is in fact the output of a technological *mixing process*, the product from which is supposed to have been *fully* mixed at the location from which increments have been extracted. The empirical evidence in Figure 17.4 is interesting in this context as it shows beyond any doubt that this objective has *not* been met—on the contrary there is solid evidence of a highly significant systematic compositional periodicity, which is an inheritance from suboptimal mixing. This is a role model type of *interpretation* of a variogram. *Were* the mixing process fully efficient there would be no residual periodicity observable in the output variogram, which would have been a flat variogram with a much lower sill.

There are many other insights to be had from proper interpretation of variograms, for example regarding the specific sill level and the magnitude of the nugget effect compared to the sill level, all to be explored in the next chapters (and in Book II).

The References list a comprehensive set of publications illustrating the features introduced above in depth. The reader is strongly recommended to consult these resources in addition; this is where it is possible to make instant, giant leaps. This list serves instead of an extensive set of summaries in conventional textbook style; here the reader gets direct access to a range of real-world examples of industrial and laboratory variographic characterisations complete with subject-matter interpretations—indispensable follow-up to the introductory subject matter above. In the few examples where the reader may not have access to the full paper, the abstracts are still valuable additions to the present chapter.

17.2 References

1. K. Engström and K.H. Esbensen, "Evaluation of sampling systems in iron ore concentrating and pelletizing processes – Quantification of Total Sampling Error (TSE) vs. process variation", *Minerals Eng.* **116,** 203–208 (2018). https://doi.org/10.1016/j.mineng.2017.07.008, 👆 bit.ly/tos17-1
2. E. Thisted and K.H. Esbensen, "Improvement practices in process industry – the link between process control, variography and measurement system analysis", *TOS forum* **Issue 7,** 20–29 (2017). https://doi.org/10.1255/tosf.97, 👆 bit.ly/tos17-2
3. E. Thisted, U. Thisted, O. Bøckman and K.H. Esbensen, "Variographic case study for designing, monitoring and optimizing industrial measurement systems – the missing link in Lean and Six Sigma", in *Proc. 8th International Conference on Sampling and Blending,* 9–11 May 2017, Perth, Australia, pp. 359–366 (2017). ISBN: 978 1 925100 56 3
4. R.C.A. Minnitt and K.H. Esbensen, "Pierre Gy's development of the Theory of Sampling: a retrospective summary with a didactic tutorial on quantitative sampling of one-dimensional lots", *TOS forum* **Issue 7,** 7–19 (2017). https://doi.org/10.1255/tosf.96, 👆 bit.ly/tos16-3
5. K.H. Esbensen, A.D. Román-Ospino, A. Sanchez and R.J. Romañach, "Adequacy and verifiability of pharmaceutical mixtures and dose units by variographic analysis (Theory of Sampling) – A call for a regulatory paradigm shift", *Int. J. Pharmaceut.* **499,** 156–174 (2016). https://doi.org/10.1016/j.ijpharm.2015.12.038, 👆 bit.ly/tos12-1
6. K.H. Esbensen and R.J. Romañach, "Proper sampling, total measurement uncertainty, variographic analysis & fit-for-purpose acceptance levels for pharmaceutical mixing monitoring", in *Proceedings of the 7th International Conference on Sampling and Blending,* 10–12 June, Bordeaux, *TOS forum*

Issue 5, 25 (2015). https://doi.org/10.1255/tosf.68, 👆 bit.ly/tos17-6

7. A. Sánchez-Paternina, A. Román-Ospino, C. Ortega-Zuñiga, B. Alvarado, K.H. Esbensen and R.J. Romañach, "When "homogeneity" is expected—Theory of Sampling in pharmaceutical manufacturing", in *Proceedings of the 7th International Conference on Sampling and Blending*, 10–12 June, Bordeaux, TOS forum **Issue 5,** 67–70 (2015). https://doi.org/10.1255/tosf.61, 👆 bit.ly/tos17-7

8. Z. Kardanpour, O.S. Jacobsen and K.H. Esbensen, "Local versus field scale heterogeneity characterization – a challenge for representative field sampling in pollution studies", *Soil* **1,** 695–705 (2015). https://doi.org/10.5194/soil-1-695-2015, 👆 bit.ly/tos17-8

9. H. Tellesbø and K.H. Esbensen, "Practical use of variography to find root causes to high variances in industrial production processes – I. Exclay (LECA)", in 6th *World Conference on Sampling and Blending (WCSB6)*, Lima, Peru, 19–22 November 2013, pp. 275–286. http://www.gecaminpublications.com/wcsb62013/, 👆 bit.ly/tos17-9

10. H. Tellesbø and K.H. Esbensen, "Practical use of variography to find root causes to high variances in industrial production processes – II. Premixed mortars", in 6th *World Conference on Sampling and Blending (WCSB6)*, Lima, Peru, 19–22 November 2013, pp. 287–294. http://www.gecaminpublications.com/wcsb62013/, 👆 bit.ly/tos17-9

11. K.H. Esbensen, C. Paoletti and P. Minkkinen, "Representative sampling of large kernel lots – I. Theory of Sampling and variographic analysis", *Trends Anal. Chem.* **32,** 154–165 (2012). https://doi.org/10.1016/j.trac.2011.09.008, 👆 bit.ly/tos8-0

12. P. Minkkinen, K.H. Esbensen and C. Paoletti, "Representative sampling of large kernel lots – II. Application to soybean sampling for GMO control", *Trends Anal. Chem.* **32,** 166–178

(2012). https://doi.org/10.1016/j.trac.2011.12.001, 👆 bit.ly/tos8-2

13. K.H. Esbensen, C. Paoletti and P. Minkkinen, "Representative sampling of large kernel lots – III. General considerations on sampling heterogeneous foods", *Trends Anal. Chem.* **32,** 179–184 (2012). https://doi.org/10.1016/j.trac.2011.12.002, 👆 bit.ly/tos8-3
14. F.F. Pitard, *Theory of Sampling and Sampling Practice*, 3rd Edn. Chapman & Hall/CRC (2019). ISBN: 9781138476486
15. P. Gy, *Sampling for Analytical Purposes*, 2nd Edn, Translated by A. Royle. John Wiley, Chichester (1999). ISBN: 978-0-471-97956-2

18 Experimental validation of a primary sampling system for iron ore pellets

We asked Karin Engström, Luossavaara Kiirunavaara AB (LKAB), Kiruna, Sweden to outline how industrial validation of a process sampling system takes place following ISO standards guidelines. These prescribe a rigorous procedure for comparison of a process sampling system with a "stopped belt" + manual sample extraction *reference method*, as a means for checking for a sampling bias, as the reference sampling system is considered to be fully TOS-compliant, i.e. representative. This chapter is a real-world example for the on-line alternative of variographic characterisation of the same iron ore pellet stream, which will follow in the next chapter. It also sets the scene for how industrial application of variographic process characterisation should always be conducted with attention to all critical prerequisites.

A̶A̲
IPE: Incorrect Preparation Error

18.1 Introduction: status of current ISO standards

Primary sampling of iron ore is well established and standardised through the International Organization for Standardization (ISO). In comparison to standardisation regarding many other mineral commodities and particulate materials, e.g. food/feed and pharmaceuticals, iron ore sampling standards are in close compliance with the TOS.[1-3] Iron ore mining and processing operations apply sampling and grade control in all parts of the production value chain, from diamond drill and blast holes all the way to process sampling of slurries, pellet feed, finished pellets and at ship loading. Sampling of iron ore is standardised through the international standard ISO 3082: "Iron ores—Sampling and sample preparation procedures".[4] The iron ore industry has improved its conformity to ISO 3082 over the last 10–20 years, especially in commercial matters. However, there are still several areas where deviations from the standard and issues with sample representativity are a reality.[3] For newly constructed sampling systems, or for in-use systems that have been modified, ISO 3082 demands *verification* of the full sampling system in accordance to ISO 3086: "Iron ores—Experimental methods for checking the bias of sampling".[5] We describe here an empirical verification and a validation experiment, as a baseline reference to be compared with an on-line variographic sampling system quality control in the following chapter.

18.2 Fundamental Sampling Principle and basic requirements for iron ore sampling systems

The Fundamental Sampling Principle (FSP) states that all parts of the lot must have *equal probability* of being selected for the sample.[6–8] This principle is equally important for primary sampling extraction as for all subsequent sampling stages, i.e. during mass reduction/sample division. ISO 3082 describes the *best location* for primary sample extraction to be at a transfer point between conveyor belts, where a full cross-section of the moving (i.e. *falling*) stream can be intercepted and extracted at regular intervals. Sampling from stationary lots, such as ships or stockpiles, is not permitted by ISO 3082, as it is *impossible* to drive a sampling device through the full lot depth and extract a complete ore column. Therefore, ISO 3082 recommends that samples are extracted **only** as the ore is being transported to or from a ship, stockpile, bunker or silo, i.e. as the ore stream is in a dynamic 1-D lot configuration.

The extraction of primary increments must comply with the following regulations to ensure that no bias is generated (well-tested TOS principles):

- a *complete* cross-section of the ore stream shall be taken when sampling from a moving stream;
- the aperture of the sample cutter shall be at least three times the nominal top size of the ore, or 30 mm for the primary sampling and 10 mm for subsequent stages, whichever is the greater;
- the speed of the sample cutter shall not exceed $0.6 \, m \, s^{-1}$, unless the cutter aperture is correspondingly increased;
- the sample cutter shall travel through the ore stream at uniform speed, both the leading and trailing edges of the cutter clearing the ore stream at the end of its traverse;
- the lips on the sample cutter shall be parallel for straight-path samplers and radial for rotary cutters; these conditions shall be maintained as the cutter lips wear;
- changes in moisture content, dust loss and sample contamination shall be avoided;
- free-fall drops shall be kept to a minimum to reduce size degradation of the ore pellets and hence minimise bias in size distribution determination;
- primary cutters shall be located as near as possible to the loading or discharging point to further minimise the effects of size degradation;
- a *complete* column of ore with nominal top size less than 1 mm shall be extracted when sampling iron ore concentrate in a wagon.

18.3 Principles and general requirements for checking sampling bias

The method for checking for sampling bias is to compare the online stream sampling system (Figure 18.1) to a reference sampling method considered to produce true and unbiased results. For 1-D process sampling, the reference is the archetype "stopped belt" sampling using a rigid sampling frame as outlined for manual increment/sample extraction, see Figure 18.2.

The number of *paired comparisons* between the reference method (method A) and the sampling method to be tested (method B) should be no less than ten. The samples A and B should be taken as close together as possible to ensure that the local, 1-D variability (heterogeneity) in the ore does not affect the bias test more than at the smallest unavoidable minimum. Quality characteristics important to the ore, such as iron content, size distribution or other metallurgical, chemical or physical properties can be used for bias testing (ISO 3082). But it is

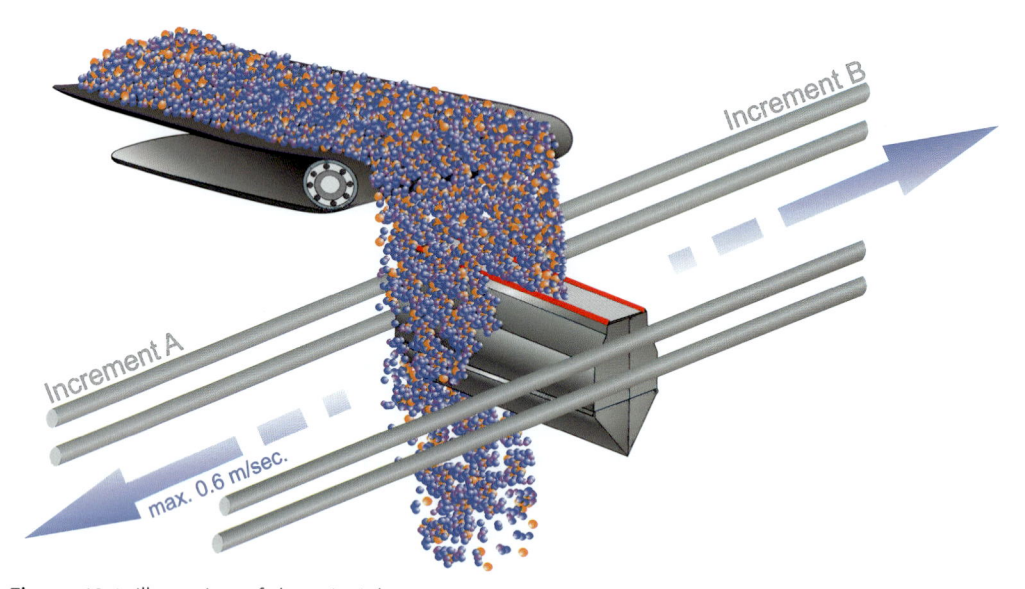

Figure 18.1. Illustration of the principles governing TOS-correct cross-stream sampling. Image credit: Martin Lischka, with permission.

Figure 18.2. Example of a "stopped belt" sampling frame allowing a complete cross-section of the ore stream to be extracted in an unbiased fashion. This will serve as a reference to the online sampling system (Figure 18.1). Image credit: LKAB.

well-known that size distribution parameters offer the most powerful check—if the size distribution of method A and B is identical, so will be the composition.

The paired measurements for the selected quality characteristics are compared using a 90% confidence interval, or an equivalent t-test, for checking if there is a bias present for the B sampling system following ISO 3086.

The current edition of both the ISO standard for checking of bias (ISO 3082) and the ISO standard for estimation of sampling precision (ISO 3085) support elimination of *outliers* identified through application of a Grubbs outlier test, taking no account whether assignable causes can be identified or not. However, this method of outlier elimination is likely to affect the bias test by favouring the tested sampling system (B) by *underestimating* the real-world, true sampling system precision. Later publications have recommended that identified outliers should only be eliminated if *bona fide* assignable causes have been identified; in addition, a new data set should also be collected and processed to ensure correct calculations.[9] (N.B. these published suggestions are in the process of being incorporated in the upcoming revisions of the ISO standards for iron ore sampling.)

18.4 Validation experiment

The experimental validation reported here was executed at a sampling system collecting the final product from an iron ore pellet plant (i.e. in-house process control sampling—**not** commercial sampling). The sampling system consists of a linear cross-stream sampler, Figure 18.1, collecting primary increments of iron ore pellets in accordance with the guidelines (ISO 3082). This system is based on a systematic time interval, collecting primary increments every five minutes. Apart from the primary sampling, the sampling system is fully automated with regard to sub-sampling division, crushing for chemical analysis, sieving analysis, abrasion index analysis and crushing strength analysis.

In the present validation experiment, the primary sampling, sub-sampling and automated sieving analysis were validated. The schematics of the sampling system is presented in Figure 18.3. The ISO 3086 approach was used with the one exception that six (not ten) paired samples were extracted.

18.5 Experimental results

The validation results for the primary linear cross-stream sampler are presented in Table 18.1. The data were analysed using a two-sided t-test with a 95% confidence level following LKAB validation rules. The t-test indicates that the primary sampler did not generate any bias for all parameters *except* for the particle size <5 mm, where a significant difference between the A and B sampling methods was identified: the linear cross-stream sampler *overestimates* the amount of *fine material* in the lot.

The reason for this discrepancy was investigated by thorough inspection of the complete sampling system, leading to the conclusion that the problem was not related to increment extraction, but rather due to *dust accumulation* on the conveyor belt during transportation of primary increments to the first rotary

Experimental validation of a primary sampling system for iron ore pellets 221

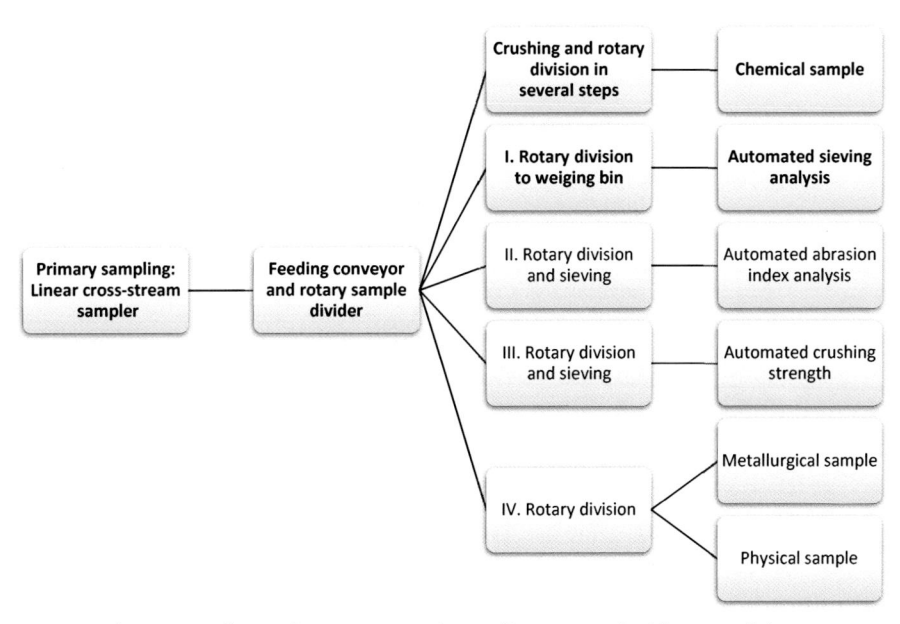

Figure 18.3. Schematics of complete automated sampling system, **bold** parts of the system were validated in the current study.

Table 18.1. Validation data for linear cross-stream sampler bias test.

	Fe	SiO$_2$	12.5–16 mm	10–12.5 mm	<5 mm
Mean: linear cross stream samples	66.69 %	2.143 %	12.5 %	59.6 %	2.0 %
Mean: stopped belt samples	66.70 %	2.135 %	12.6 %	59.4 %	1.2 %
Mean difference	−0.01 %	0.008 %	−0.1 %	0.2 %	0.8 %
Standard deviation for mean difference	0.043 %	0.026 %	1.6 %	1.8 %	0.4 %
Critical t-value	2.57	2.57	2.57	2.57	2.57
Statistical t-value	−0.67	0.71	0.13	−0.32	−5.0
Significant difference?	No	No	No	No	Yes

sample divider. This problem was *counteracted* by improving ventilation to decrease the amount of "ambient dust" in the environment around the primary sampler, and by improving the physical shielding of the conveyor, ensuring no dust can reach the belt from outside sources.

To ensure TOS-correctness (un-biasness) of the automated sieve analysis connected to the primary sampling system, a bias test was also conducted in relation to a well-controlled reference laboratory sieving setup. The results of the t-test for the six compared samples are presented in Table 18.2. This t-test does not show any statistical difference between the two pieces of sieving equipment and the validation of the automated sieve is, therefore, approved.

18.6   Discussion

The result from the validation experiment of the primary linear cross-stream sampler illustrates the critical importance of validation and bias testing. Even though the sampling system was constructed according to TOS principles and ISO 3082, a bias *could* in fact be detected, due to sample *contamination* after primary sample extraction, revealing an Incorrect Preparation Error (IPE) effect.

Table 18.2. Validation data for the automated sieve analysis sampler.

	12.5–16 mm	9–12.5 mm	5–9 mm
Mean: automated sieve	18.7 %	74.9 %	3.5 %
Mean: reference sieve	18.5 %	75.2 %	3.9 %
Mean difference	0.2 %	−0.3 %	−0.4 %
Standard deviation for mean difference	4.7 %	3.9 %	0.9 %
Critical t-value	2.57	2.57	2.57
Statistical t-value	0.09	−0.19	−1.07
Significant difference?	No	No	No

The decision to decrease the number of paired samples from ten to six in this validation experiment was due to cost considerations at the discretion of LKAB management. This provides for a "smart" option; the experiment had the possibility to be extended with the four additional paired samples **if** the first results were inconclusive.

Even though a base-line validation of all new or modified sampling systems is required (ISO 3082), *continuous* monitoring and control is also mandatory—to ensure continuous representative sampling and analytical results.

Apart from regular maintenance and visual inspection, continuous variographic characterisation is an efficient way of monitoring and quality grading process sampling systems over time, ensuring against that significant deviations in sampling or analytical variability has accidentally been introduced. The variographic approach is well described in previous chapters, and will be applied in the next chapter to the same parameters as were used here.

Multi-increment sampling
This LKAB deviation from a canonical number of samples is an example of the "multi-increment sampling" approach in which the full number of required samples are sampled in the field (the plant in this case), but only a reduced set are analysed at first. If necessary, the complementary samples can be analysed in a second campaign without loss of the full experimental design advantages.

18.7   References

1. P. Gy, "Sampling of discrete materials—a new introduction to the theory of sampling I. Qualitative approach", *Chemometr. Intell. Lab. Syst.* **74**, 7–24 (2004). https://doi.org/10.1016/S0169-7439(04)00167-4, bit.ly/tos18-1
2. R.J. Holmes and G.J. Robinson, "Codifying the principles of sampling into minerals standards", *Chemometr. Intell. Lab. Syst.* **74**, 231–236 (2004). https://doi.org/10.1016/j.chemolab.2004.03.011, bit.ly/tos18-2
3. R.J. Holmes, "Common pitfalls in sampling iron ore", *Proceedings 8th World Conference on Sampling and Blending*, 9–11 May, Perth, Australia, pp. 261–264 (2017).

4. International Organization for Standardization, *ISO 3082: Iron Ores—Sampling and Sample Preparation Procedures* (2009).
5. International Organization for Standardization, *ISO 3086: Iron Ores—Experimental Methods for Checking the Bias of Sampling* (2006).
6. P. Gy, *Sampling of Particulate Materials—Theory and Practice*, 2nd Edn. Elsevier, Amsterdam (1979).
7. F.F. Pitard, *Pierre Gy's Sampling Theory and Sampling Practice*, 2nd Edn. CRC Press Inc., Florida (1993).
8. J.E. Everett, T.J. Howard and B.J. Beven, "Precision analysis of iron ore sampling preparation and measurement overcoming deficiencies in current standard ISO 3085", Min. Technol. **120,** 65–73 (2011). https://doi.org/10.1179/174328631 1Y.0000000002, bit.ly/tos18-8
9. DS 3077. *Representative Sampling—Horizontal Standard*. Danish Standards (2013). http://www.ds.dk, bit.ly/tos16-1

19 Industrial variographic analysis for continuous sampling system validation

19.1 Variographic analysis

Variographic process data analysis was presented in the preceding chapters, for example, References 1–6. This chapter exclusively addresses application in the industrial domain, particularly regarding the possibility to conduct continuous on-line measurement system control. The key issue is that the total measurement system variability is composed of both the Total Sampling Error (TSE) *plus* the Total Analytical Error (TAE) contributions.

TAE: Total Analytical Error
TSE: Total Sampling Error
SSA: Specific Surface Area
FSP: Fundamental Sampling Principle

19.2 Continuous control of sampling systems

Validation of newly installed, recently upgraded or modified industrial sampling systems is essential for reliable process monitoring. This ensures documentation of representative sampling and reliable analytical results. Validation of process sampling systems has traditionally been based on carefully extracted physical samples. Sometimes this approach has not been thought through fully, however, but has unwittingly led to employing a different sampling routine for process measurement system quality control to what is used in routine operations. This opens up all manner of IDE, IEE a.o. which will invade and adversely

Most importantly, sequential "on-line" variographics allow this critical QC/QA to be based on routine process samples extracted under routine conditions.

In this chapter a wide variety of "analytes" are subjected to variographic data analysis with detailed industrial interpretations, both physical material characteristics as well as intrinsic and extrinsic chemical parameters.

influence the validation protocol. This setup is obviously to be avoided at all costs!

However, for *continuous control* of existing sampling systems it is equally important to ascertain that representativity is not infringed upon due to undetected changes in the sampling or process systems (trends, periodicity, equipment drift, upsets or component replacement etc.). Furthermore, continuous control can ensure that the measurement system still functions satisfactorily if/when significant material compositional changes are introduced. Continuous variographic characterisation is a powerful tool for inspecting the *sum* of all the components contributing to the observable process variability over time. Variographic analysis can identify and quantify, as well as distinguish *between*, sampling, analysis and process variabilities; we explain how below. Variographic analysis can *decompose* the total, *apparent* process variability. This approach can, therefore, be used to *detect* changes in any source of variability, no matter if it originates from the process or from the sampling system.

This chapter presents a selected set of examples where variographic analysis has been applied to operating sampling systems at the LKAB concentrating and pelletising plants in Sweden, here serving as inspirational *generic examples* of the amazing amount of information that can be derived through the use of variographic analysis. The importance of this approach to process characterisation is actually much wider than industrial processing and manufacturing; this is a most interesting area for variographic application (see a.o. Book II).

19.3   9–12.5 mm size fraction of iron ore pellets

Size analysis of iron ore pellets is performed by the sampling system validated in the previous chapter, a linear cross stream sampler collecting primary samples every five minutes. The

Industrial variographic analysis for continuous sampling system validation

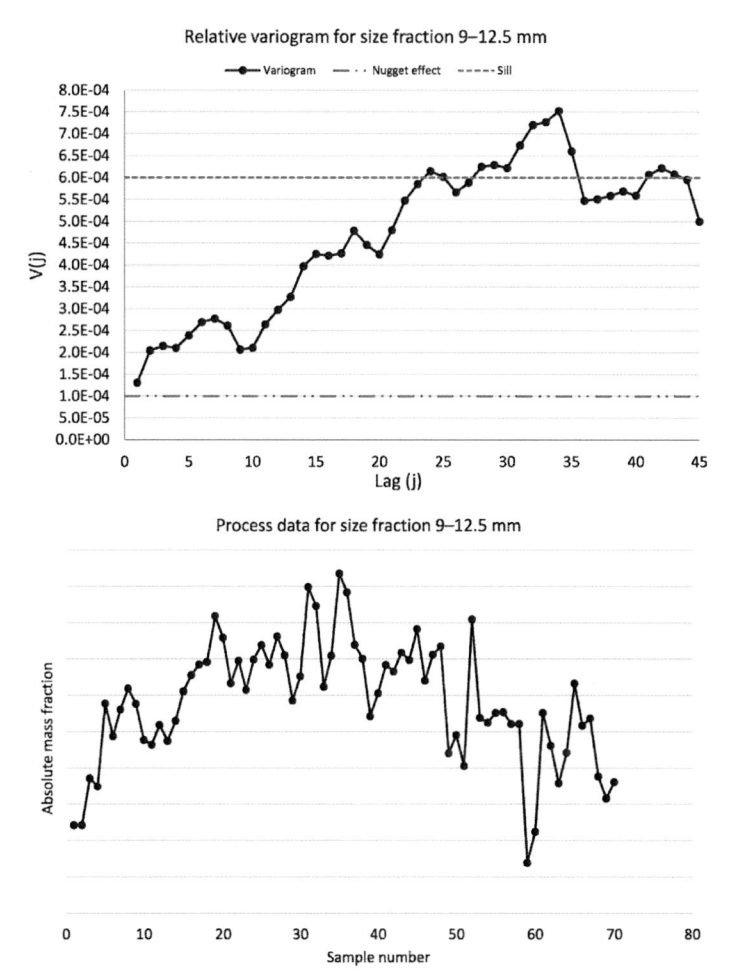

Figure 19.1. Variogram and process data for the 9–12.5 mm size fraction of iron ore pellets. Sill, dashed line; nugget effect, dotted and dashed line. The nugget effect to sill ratio equals approximately 15 %.

complete sampling system is automated with subsequent rotary splitting, sub-sampling and automatic size analysis, presenting results approximately every hour. The variogram and process data for the 9–12.5 mm size fraction is presented in Figure 19.1. Successive variograms present similar results over several time periods, indicating similar sampling and analytical variability as was observed during the validation experiment.

Variograms must always be presented together with the raw process data from which they are derived. The interpretation of the size fraction variogram is particularly easy and clear in the present case. As would be expected from general TOS experiences, a well-calibrated and maintained linear cross stream sampling system *should* give a reliable picture of the true process variation. This can be seen by the fact that the nugget effect to sill ratio is only approximately 15%, indicating that the total sampling and analysis system does indeed allow an accurate reflection of the true process variation at any one time. This on-line sampling system is validated and can be continuously verified as *fit-for-purpose representative*.

Nugget effect-to-sill ratio
The fraction of the sill level (the maximum variability in the data series) made up of the nugget effect, the "Nugget effect-to-sill ratio" is a highly informative measure of the uncertainty stemming from the compound process measurement system. Usage of this important facility is demonstrated in several of the references to this chapter, particularly References 5 and 6. In general, this ratio should be below 30%, see DS 3077 (http://www.ds.dk) and many other references.

19.4 Specific surface area of magnetite slurry

The specific surface area (SSA) of magnetite slurry is an important parameter for the process of balling iron ore pellets. The surface area is determined in the milling of the ore and is used for feedback process control of the same milling process. The sampling system for SSA employs a continuous primary system known as a *shark fin sampler* in concentrating plant 3, and spear collection of a continuous flow from a pressurised pipe in concentrating plant 2. Both these sampling systems violate the TOS' Fundamental Sampling Principle (FSP) by not acquiring a complete cross-section of the slurry stream, but only collecting a part of the stream all of the time. Whether this impairs proper process control in practice is, amongst other things, related to the degree of heterogeneity of the flowing material; well-mixed material *might* be sampled fit-for-purpose with this approach, while significantly heterogeneous material will surely fare badly! This is the "gospel" according to the TOS; witnessed by the TOS literature.[1-6]

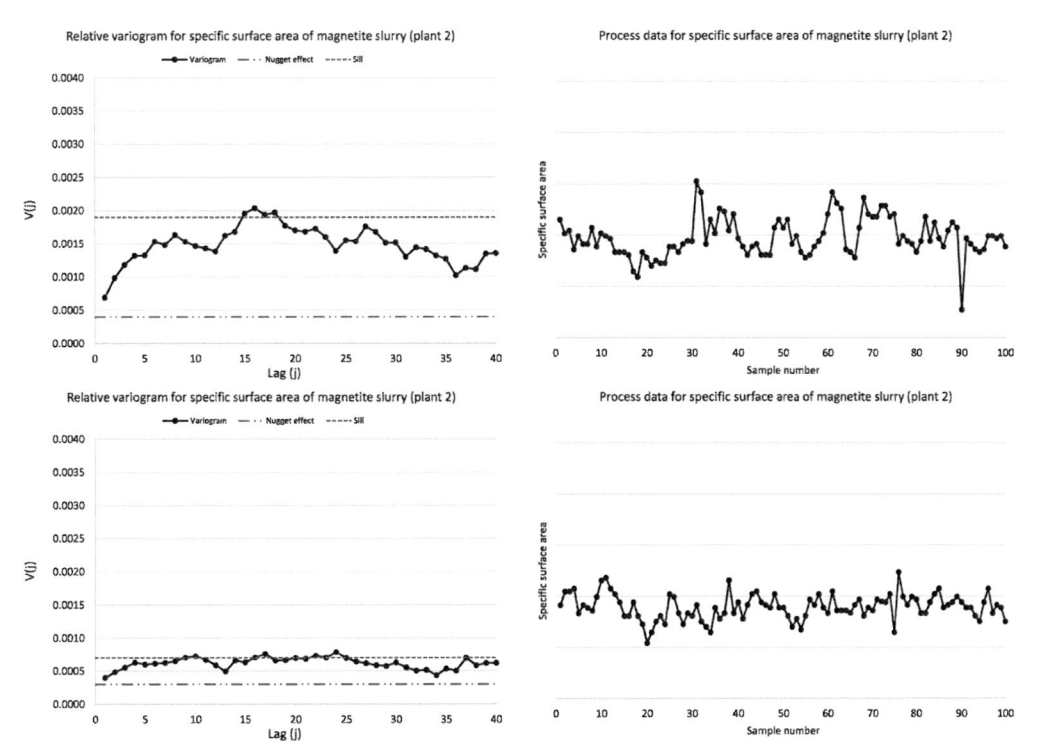

Figure 19.2. Variograms and process data for the SSA sampling system in plant 2 (TOS-correct secondary sampling). The two variograms are from different time periods, showing similar nugget effects but significantly different sill levels (see text). Sill, dashed line; nugget effect, dotted and dashed line.

The secondary sampling for SSA in plant 3 consists of a *grab sample* collected from the primary sample stream every four hours; also much maligned by the TOS. However, the secondary sampling for SSA in plant 2 is a TOS-correct increment sampling, collecting and combining several increments to form a composite sample accessed for analysis every four hours. The samples from both plants are analysed according to the Blaine method using an automatic analyser.

The variograms for SSA sampling systems in plants 2 and 3 show similar nugget effects (absolute magnitudes) in all produced variograms. However, the sill levels change significantly

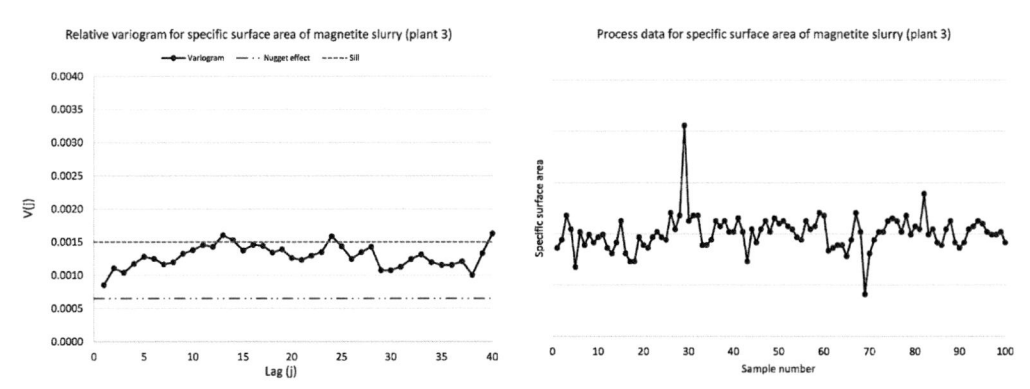

Figure 19.3. Variograms and process data for the SSA sampling system in plant 3 (secondary grab sampling, TOS-incorrect). Sill, dashed line; nugget effect, dotted and dashed line. The variograms nevertheless show similar nugget effect as plant 2.

over the different time periods analysed. Three typical (varying) variograms are presented in Figures 19.2 and 19.3.

The first variogram for plant 2 reveals a satisfactory performance for the total measurement system, with a nugget effect to sill ratio of approximately 20%, allowing a fair picture of the true process variations. In the second variogram for plant 2, the nugget to sill ratio is approximately 50%, which could lead to the conclusion that the uncertainty of the total measurement system is too high. But in this case the reason for the high nugget to sill ratio is rather a stable process showing a very small overall variation, thus lowering the sill and therefore *inflating* the nugget to sill ratio.

The process being monitored, and the attendant on-line quality check of the total measurement system presented by the variogram in Figure 19.3, is characterised by the fact that the process is dominantly stable, with only a few, isolated deviations. Process stability leads to a well-defined *low sill*, which in itself is important information for process operators. Low, stable sills carry the main message for such cases: currently the process does not need *active* controlling measures. The measurement system also performs well here, even if the nugget effect to sill ratio *appears* to be as high as 45%. In situations where the sill is low, one need not put emphasis on the measurement system quality index, nugget-to-sill ratio, since it is artificially inflated by the low sill.

19.5 Iron grade in magnetite slurry

Sampling for iron grade in magnetite slurry is carried out by a completely automated sampling system. The primary sample is also collected by a shark fin sampler, which collects only part of

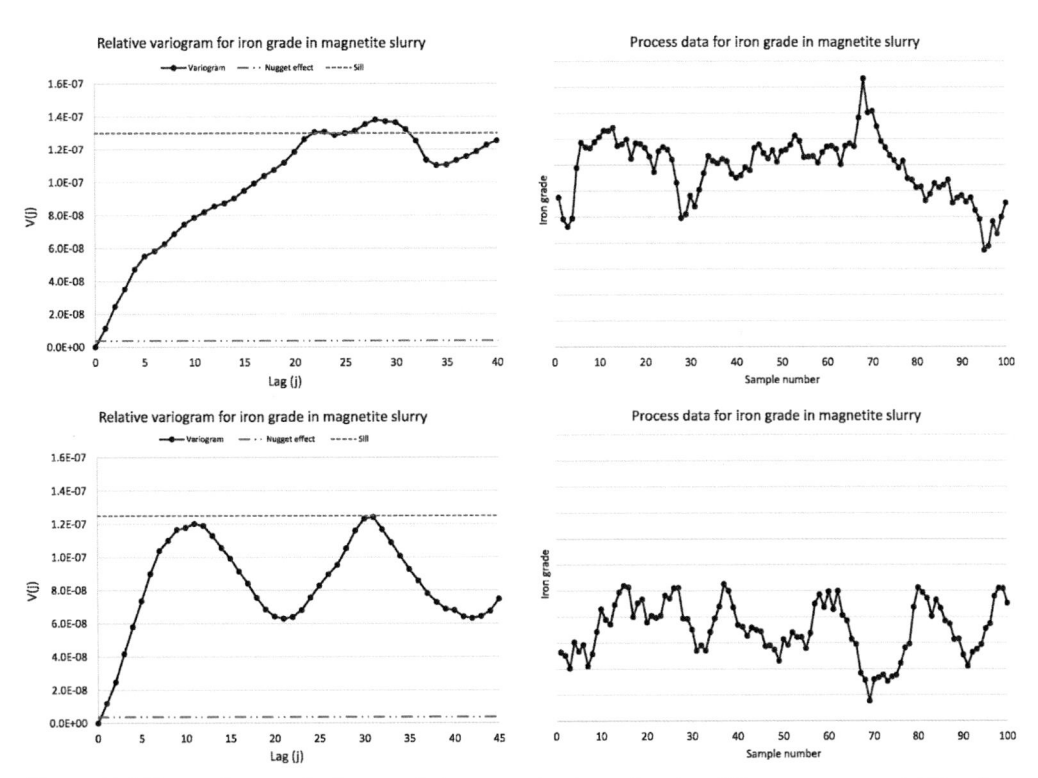

Figure 19.4. Two variograms for iron grade in magnetite slurry showing different process variability characteristics at different time periods. For a few of the analysed time periods, a clear cyclic behaviour could be identified, see text for detailed interpretation.[5] Note the *excellent* process measurement system characteristics, witnessed by a nugget effect-to-sill ratio of ~5% only.

the stream all of the time. The sample is subsequently split and dried before being sent to a robotic X-ray fluorescence laboratory for analysis of Fe, Si, P, V and several other important parameters. Two variograms for this sampling system are presented in Figure 19.4, representing different time periods of the continuous process operation (these data were previously also published in Reference 5).

Both these variograms show similar nugget effects and similar sill levels. There is, however, one very clear difference as the second variogram shows a clear periodicity of approximately 20 lags corresponding to 8–10 h. Only shorter, isolated time periods exhibit periodicity for this sampling system, but through a continuous variographic characterisation they can be easily detected. Close familiarity with the details of this process makes it likely that periodicity, when detected, is due to fluctuations in the milling capacity, which affects the floatation efficiency and, thereby, also the iron grade. In some cases, periodicity can also be derived from the sampling system, but this is not considered the case here. Either kind of periodicity is important to recognise and avoid, lest unnecessary and exaggerated process changes *might* be introduced due to over-corrections by ardent process operators. By applying continuous on-line variographic analysis, it is possible to identify and characterise *diverse* causes for deviations from stable process conditions. This is possible due to the well-controlled total measurement system characterisations with nugget-to-sill ratios down to approximately 5 %.

19.6 Conclusions

Variographic analysis will in most cases allow meaningful *decomposition* of observed (raw) process data variabilities into contributions stemming from TSE and/or TAE, revealing the true process variability. The important exception is the case of a *low sill*, which need not lead to any concern on behalf of the variogram data analyst, as this case is signalling that all is well (stable) with the process being monitored.

Three typical variogram appearances with different nugget effect to sill ratio characteristics were presented. There are no cases in practical variographic analysis other than either a *low sill* (in which case all is good due to low process variability) or a significantly high(er) sill (also *all good*, but here specifically with respect to the possibility of variographic decomposition illustrated above).

Since the first version of this chapter that appeared as a column in *Spectroscopy Europe*, Karin Engström has successfully defended her PhD thesis at Alborg University, campus Esbjerg (December 2018).[6] The thesis is freely available, and complements this chapter massively with respect to practical variographics in industry and measurement technology. References 1–6 constitute a unique, comprehensive base from which to learn much more about variographics.

https://www.spectroscopyeurope.com/sampling/experimental-validation-primary-sampling-system-iron-ore-pellets, bit.ly/tos19-a

K. Engström, *Sampling in Iron Ore Operations: Evaluation and Optimisation of Sampling Systems to Reduce Total Measurement Variability*. PhD Thesis, Aalborg Universitetsforlag (2018). https://doi.org/10.5278/vbn.phd.eng.00068, bit.ly/tos19-6

19.7 Acknowledgements

The authors express their gratitude to LKAB for allowing this variographic study of current process data and for publishing of the results.

19.8 References

1. R. Minnitt and K.H. Esbensen, "Pierre Gy's development of the Theory of Sampling: a retrospective summary with a didactic tutorial on quantitative sampling of one-dimensional lots", *TOS Forum* **Issue 7,** 7–19 (2017). https://doi.org/10.1255/tosf.96, bit.ly/tos16-3

2. K.H. Esbensen and P. Paasch-Mortensen, "Process Sampling: Theory of Sampling – the Missing Link in Process Analytical Technologies (PAT)", in *Process Analytical Technology*, Ed by K. Bakeev. Wiley-Blackwell, pp. 37–80 (2010). https://doi.org/10.1002/9780470689592.ch3, 👆 bit.ly/tos19-2
3. R. Minnitt and F.F. Pitard, "Application of variography to the control of species in material process streams", *J. S. Afr. Inst. Min. Metall.* **108(2),** 109–122 (2008).
4. F.F. Pitard, *Theory of Sampling and Sampling Practice*, 3rd Edn. Chapman and Hall/CRC (2019). ISBN 9781138476486
5. K. Engström and K.H. Esbensen, "Evaluation of sampling systems in iron ore concentrating and pelletizing processes - Quantification of Total Sampling Error (TSE) vs. apparent process variation", *Proceedings of MEI Process Mineralogy 16*, Cape Town (2016).
6. K. Engström, *Sampling in Iron Ore Operations: Evaluation and Optimisation of Sampling Systems to Reduce Total Measurement Variability*. PhD Thesis, Aalborg University (2018). https://doi.org/10.5278/vbn.phd.eng.00068, 👆 bit.ly/tos19-6

20 Theory of Sampling (TOS): *pro et contra*

Sampling takes place in everybody's daily life. Consciously or unconsciously, we all take decisions regarding how to select and collect the things we need or are interested in, be this vegetables, meat or coffee in the supermarket, wine or other beverages… or "samples" for academic research projects or in industry—and everything in between. Those who have been curious enough to reflect on everyday decision-making processes discovered immediately that sampling decisions often make all the difference. This is why an incipient theory started to be elaborated. One individual, more brilliant than others, made a giant step forward in the evolutionary thinking on sampling and developed what became *the* Theory of Sampling (TOS); his name was Pierre Gy. Another chapter in this book is dedicated as a tribute to his life's monumental scientific achievements, which started in the year 1950.

Below, with the experience gained over 70 years, we discuss cases both *pro et contra* proper sampling. Readers of earlier chapters will readily understand why cases *pro* the TOS can be marshalled, but will rightly wonder: *why* cases *contra* the TOS?

Well, it is time to stray a bit outside the strictly scientific and technical issues and make an attempt to understand *why* TOS is still not universally accepted despite being universally applicable. Below is presented and discussed various motivations for taking on, or not, representative sampling, or embarking on a project replacing existing sampling systems that have been found faulty and non-representative. The very first chapter in this book framed the key issue squarely: "What is the meaning of analysing a demonstrably non-representative sample?" At the time the

AA
TAE: Total Analytical Error
TSE: Total Sampling Error
ISE: Incorrect Sampling Errors
IDE: Increment Delineation Error
IEE: Increment Extraction Error

> There are no made up arguments in this chapter, believe it or not. There is, of course, neither reason nor justification for attribution, but should individual readers recognise themselves below, so much the better. They are due a big Thank You for contributing to the ever increasing dissemination of proper TOS.

conclusion was straightforward: There is no meaning! In spite of this impeccable logic, proponents of the TOS still often meet arguments (of bewilderingly different sorts) why one should *not* involve the TOS. It is illuminating to understand *what* are the drivers and arguments behind this surprising attitude. Stepping into what we think are the wrong shoes, trying to understand the fundamental reasons for the existence of the at times strong resistance to the TOS is necessary, even if not sufficient on its own, to find a more effective and successful communication strategy to explain that sampling and representativeness are two sides of the very same coin. Hence cases *pro et contra* the TOS are presented and discussed below.

20.1 A powerful case for the TOS in trade and commerce

According to international trade agreements and codes, disputes between *buyer* and *seller* are to be pre-empted by duplication (or triplication) of primary samples, of which one is analysed by the buyer, the other by the seller, while a third sample is archived to be used *if* disputes can only be resolved in a court of law. Sometimes a third party is called for who then analyses the archival sample, or (much less frequently), is asked to perform a completely new primary sampling + analysis. Usually, however, only the two analytical results from the buyer and seller are compared, and *should* ideally fall within a commonly agreed upon uncertainty interval, specified in the contract; the simple average value is then often used for the pertinent business purposes, see Figure 20.1.

The interesting case is when analytical differences *exceed* this acceptance interval, in which case the trade codes *mandate* that the archival sample is forwarded to and analysed by a third, *independent* party, whose analytical result is sometimes used directly by *fiat*. If this is not acceptable to one or both parties, the dispute

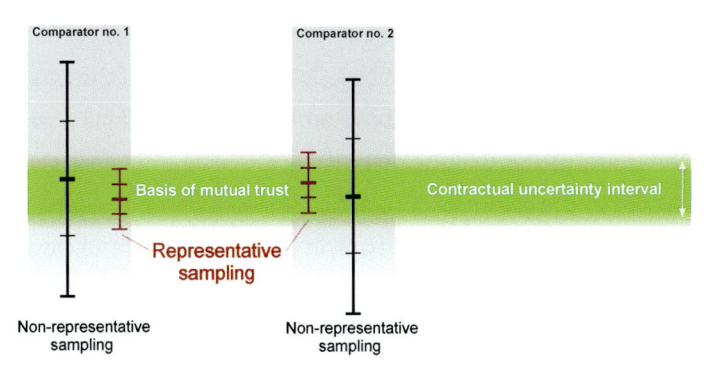

Figure 20.1. The consequences of non-representative sampling are identical for buyer and seller—an inflated sampling variability (black and denoted "Non-representative sampling") making it very difficult to be able to satisfy the contractual uncertainty interval (green). Things get completely out of control when both buyer and seller, and even a third arbitration party, may freely choose their own sampling procedure. Resolution of the analytical result comparison issue is only possible when all parties agree **only** to use representative sampling procedures (red/"Representative sampling").

goes to arbitration in a court of law. The court will, in most cases, then *dictate* to use the average between the two *nearest* of the three analytical values, upon which to conduct the salient business transaction. This arbitration approach *appears* logical and is indeed easy to follow, and is never questioned further—likely because there is always a guaranteed resolution.

However, there is a hidden elephant in the room!

There are very rarely sufficient stipulations on *how* the primary samples are to be extracted. Indeed, it is very often tacitly understood that each party or stakeholder is free to use whatever sampling procedure is preferred. The focus is overwhelmingly on the magnitude and the quality of the final analytical results. It is thus acceptable that the seller and the buyer may wish to perform sampling independently, for example having the seller sample at the port of loading of a ship's cargo, while the buyer samples the same cargo but upon arrival at the receiving port. This is because every pair, or every triplicate set, of primary samples is simply *assumed* to be fully representative of the cargo in question; otherwise the above arbitration rules will fall apart. The crucial issue is that the dominating sampling error effects are *invisible* in the gamut of contractual stipulations—it is all about the analytical results, and perhaps, in the more thoughtful cases, also about the quality of the analytical determinations involved (focus on TAE *instead* of TSE + TAE).

Sampling procedures for which the TOS demands elimination of all bias-generating errors are not heeded (Incorrect Sampling Errors, ISE) will lead to *biased sampling*. This leaves everybody without control of the magnitude of the influence of the material heterogeneity. This will unavoidably lead to a significant *inflation* of the practical sampling variability, the more so the higher the lot heterogeneity. When one, or both, parties in an analytical dispute are *not* in compliance with the prerequisites for representative sampling, the empirical sampling variability is highly likely to be much larger than the commonly agreed upon contractual uncertainty interval (dependent on the material heterogeneity). This translates directly into a high probability that the analytical results from both parties cannot be resolved, but will have to go to arbitration. This is the *status quo* for very many current international trade agreements, codes and contracts. The degree to which this *scheme* results in the need for arbitration is directly proportional to the inherent heterogeneity of the material involved, and to the degree of deviation from the principles that guarantee representative sampling. Lots with low heterogeneity will rarely experience a need for arbitration, but if/as heterogeneity goes up, so will the number of cases in which resolvable analytical results are not possible.

The key feature here is that the degree of heterogeneity of the lot or material, as sampled by the specific procedure in use (representative or not), is the real determinant w.r.t. the magnitude of the difference between analytical results—and most emphatically *not* the aptitude of the analytical laboratories involved. Usually analytical errors are very well under control (TAE)—indeed these are practically always negligibly small compared to the dominant total sampling error effects (TSE). This all means that there will never be a bona fide *common basis* upon which to evaluate the magnitude and the significance of the difference between any two, or three, analytical results in *all* resolution efforts. As long as there is no agreement or contract that legally *demands* representative sampling, there will never be an objective basis, nor a rational treatment of analytical disputes. This is unfortunately the *status quo* in very many current cases.

For both buyer and/or seller the consequence of non-representative, i.e. biased sampling is a completely unnecessary *inflated* sampling variability, compared to unbiased procedures, Figure 20.1. Only representative procedures are able to deliver a minimum sampling uncertainty that can be compared to the contractual uncertainty interval. Things get really out of control if/when buyer and seller, and/or an arbitration agency, can freely choose their own sampling procedure: see for example Figure 20.2. There is only one way out of this hidden issue not comprehensively recognised in current trade agreements and codes—all sampling procedures involved *must* be representative, i.e. compliant with the TOS, for example with reference to the standard DS 3077.

The mind boggles when it is realised that a single paragraph is able to rectify the fatal quagmire outlined above, a paragraph that just needs to be included in all contracts forthwith when issues of sampling are on the agenda:

"All sampling procedures invoked to secure primary samples (as well as all sub-sampling operations needed to produce the analytical aliquot), whether by buyer, seller or an arbitration

> *"All sampling procedures invoked to secure primary samples (as well as all sub-sampling operations needed to produce the analytical aliquot), whether by buyer, seller or an arbitration agency, shall be compliant with the principles of representative sampling as laid out by the Theory of Sampling (TOS), as codified in the standard DS 3077 (2013). All sampling procedures involved must be adequately and fully documented."*

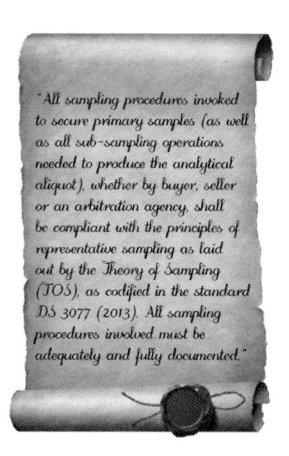

20.2 Cases against the TOS (science, technology, commerce, trade)

What follows tries to cover the scene of reluctance, unwillingness or downright ignorance w.r.t. invoking the TOS experienced over the last 20 years. Unbelievingly, the following statements are **true** (comments and directions by the present author in **boldface parenthesis**).

"You claim that this sampler is not representative, based on a Replication Experiment characterisation—but this sampler has been in use for over 30 years—how *could* it be wrong?" [**The Replication Experiment (RE) has not been understood—training, or re-training, is critical.**]

"I *cannot* tell my customers that these samplers, which we have installed in our plants for decades, now "suddenly" are wrong, and need to be replaced!" [**This sales person clearly has a very different agenda than being responsible for selling the customer a system that a.o. guarantees representative analytical results—this sales person must be (re-)trained regarding the importance of representativeness and the economic consequences of faulty decisions made by the customers. What is the meaning of analysing demonstrably non-representative samples? There is none!**]

"The sampling standard must be easy to follow, and to implement—or else it will not be used." [**A score of similar "simplicity arguments" against invoking the TOS have been overheard in numerous standardisation committees and technical task forces, all the more incredible as such are *supposed* to be

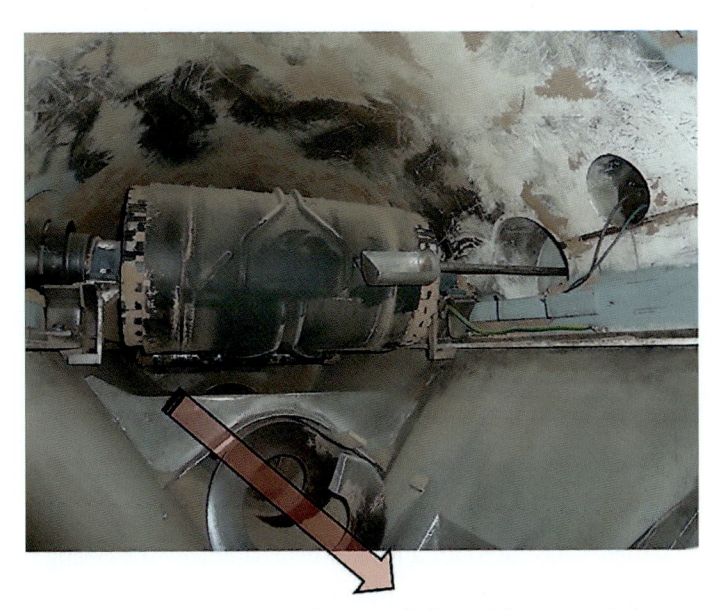

Figure 20.2. An attempt to design a dedicated "conveyor belt sampler", intended to sample from a falling stream of particulate matter. Conveyor belt (top), continuous screw feeder (bottom). In between, a hopelessly inadequate manually operated scoop sampler incurring significant IDE (Increment Delineation Errors) as well as grave IEE (Increment Extraction Errors). This scoop is totally unable to supply a continuous cross-slice of the moving stream of matter, and will manifestly also be subject to severe overflow. In spite of a critical sampling audit, this sampler is still in use because "there is no room to install a replacement sampler".

staffed by the most knowledgeable and experienced experts. Crucially in this context: What is the meaning of analysing a demonstrably non-representative sample? There is none!]

"The TOS principles are not required in this specific ISO standard." [Many sections exist in a plethora of current standards ostensibly dealing with "sampling", but sadly with very little, or no, cognisance of the TOS. Ignorance of the law is no excuse, however. Figures 20.2 and 20.3 are vivid illustrations of blatant disregard of this kind, on the altar of "practicality".]

The client will not accept/not pay for such overly complex samplers! I will not meet my quota if insisting on invoking this expensive TOS equipment". [It is the responsibility of the sales force to be competent w.r.t. the TOS to such a degree so as to be able to *explain* the consequences of buying a, say $1M processing plant (this is a true example, not an exaggeration) while insisting on installing demonstrably non-representative sampling equipment. What is the meaning of analysing demonstrably non-representative samples? There is none! It is the responsibility of the pertinent sales force supervisors to ensure that front-line sales personal have adequate TOS skills.]

"There is no room for a replacement sampler—the ceiling is too low... It is prohibitively costly to raise the roof on the building." [Let these statements be placeholders for a whole range of similar "practical arguments" why a representative sampler simply cannot be considered. Clearly economics goes before representativity here—but what is the meaning of analysing a demonstrably non-representative sample? There is none!]

"Tradition has always been *not* to cut a full slice of the stream of matter across the whole width of the flow. There is simply not space around the conveyor belt to allow the sampler to reach all the way across." [Any responsible person involved must be able to explain the fatal consequences of allowing incorrect sampling errors to influence the sampling process, and indeed of using a manual sampling process. Practicality, perceived technical difficulty, economics, logistics... and a host of other contrarian arguments cannot be the *driver* behind primary sampling—only representativity can! What is the meaning of analysing a demonstrably non-representative sample? There is none!]

Figure 20.3. Scene from an analytical laboratory (sample preparation front room). The full content of the white plastic container (left), a carefully collected composite sample (12 kg), is in the process of being coarse-crushed in a study aimed at illustrating the value of proper TOS procedures in all steps of the complete "from-field-to-aliquot" pathway. Unbelievably, an authoritative laboratory stakeholder (the laboratory head in fact) suggested, with conviction and strength: "It is not necessary to go to such lengths as to crush all this material (for each field sample)—just pick and crush a 'suitable fragment' of the mass commensurate with the subsequent fine-crushing capacity (20 g)—this saves a lot of work!" This statement exposes the wide-spread work efficiency argument against invoking the TOS principles to the extreme, combined with blatant neglect of the main characterisation of lot materials (rocks in this case): *heterogeneity*—alas with catastrophic results in the form of a 1/600 reduction of the coverage of the significantly heterogeneous material. "Luckily" (in fact not by luck alone...), the laboratory student assistants involved were well trained w.r.t. the TOS and rejected this proposal firmly.

"Do you really think that Gy was the only one to understand sampling? You guys have been *brain-washed*. Sampling can be carried out in many ways, and they all work". **[The first part of this statement is just brash and personal, showing a complete lack of understanding of the TOS; we leave it on the irrelevant scrap heap of personal insults. The second part represents a complete lack of demonstrable evidence—while on the contrary the TOS community has always gone out of its way to show why "other theories of sampling" in fact**

> **Only one Theory of Sampling (TOS)**
> Comprehensive and accountable documentation for the claim that there is in fact only one TOS is presented in 👆 bit.ly/tos2-1a

do not lead to representativity; witness, for example, all the literature cited in this book.]

"Linking a discipline to a person is scientifically wrong—a plurality of views is a key value in science. [At first view this "argument" would appear to contain a substantive general warning—but it is in fact a gross misrepresentation in the case of the TOS. It takes an in-depth discussion to treat this statement with the seriousness it deserves, which is left to a later occasion, but see Chapter 6 dedicated to the scientific achievements of Pierre Gy.]

"Applying the TOS costs too much—the current sampling protocol has been in use for 20 years and there has never been any complaint! We are not going to change everything today without solid economic evidence. [A plethora of solid evidence of economic losses due to faulty sampling exists in the TOS literature, see also Chapter 21. Any TOS-competent proponent is able to run a Replication Experiment (RE) or a variographic experiment, estimating the effective sampling variability, given availability of relevant data. From this, the economic gains or losses are easily calculated.]

"Have you ever seen a shipment of grain or soy beans? Do you really mean that we are to *ensure* that *every single grain* has the same probability of being sampled?" [The individual making this statement is unaware of the critical role played by the Fundamental Sampling Principle (FSP). The key TOS tenet is only to sample where ship cargoes are in an effective 1-D transportation state, e.g. in a grain elevator, in a pipeline or on a conveyor belt, see earlier chapters.]

Many more, similarly illustrative statements can be marshalled, all in the same tune: it would appear that pretty much anything *but* representativity is considered acceptable drivers for sampling! Obviously, if the reader has read the book up until this

> *Que faire?*
>
> Education, outreach, TOS courses, didactic and convincing scientific and technological publications, guest appearances at symposia, conferences, annual meetings in ever more diverse application fields in science, technology, trade and industry! And above all, clear, simple-to-grasp case histories that demonstrate not only the hidden economic losses following ignorance or unwillingness to introduce the TOS! It is hoped that at least some intrepid readers will feel compelled to start contributing, and also to collaborate in this quest!

point there will be universal agreement that this is the state-of-affairs cannot be more wrong, however.

The last 15–20 years has seen an explosion of achievements within all of the above areas, at all levels from the very first awareness of sampling issues… to whole textbooks. But less will also do, even much less. Below follows a selection of easily available sources from which to educate oneself in all matters re. the TOS (all of which also refer to more in-depth sources). Or what about seeing this section as a scientific trove of presents?

20.3 Important reading with which to catch the attention of newcomers to the TOS

DS 3077. *Representative Sampling—Horizontal Standard*. Danish Standards (2013). http://www.ds.dk, bit.ly/tos16-1

Sampling Columns in *Spectroscopy Europe*; see complete list at https://www.spectroscopyeurope.com/sampling, bit.ly/tos20-2

Proceedings of the World Conference on Sampling and Blending series: https://intsamp.org/proceedings/, bit.ly/tos20-3

Issues of the sampling community's newsletter, *TOS forum*: https://www.impopen.com/tosf, bit.ly/tos20-4

K.H. Esbensen, R.J. Romañach and A.D. Román-Ospino, "Theory of Sampling (TOS): A Necessary and Sufficient Guarantee for Reliable Multivariate Data Analysis", in

Pharmaceutical Manufacturing in Multivariate Analysis in the Pharmaceutical Industry, Ed by A.P. Ferreira, J.C. Menezes and M. Tobyn. Academic Press, Ch. 4, pp. 53–91 (2018). ISBN 978-0-12-811065-2

R.C.A. Minnitt and K.H. Esbensen, "Pierre Gy's development of the Theory of Sampling: a retrospective summary with a didactic tutorial on quantitative sampling of one-dimensional lots", *TOS Forum* **Issue 7,** 7–19 (2017). https://doi.org/10.1255/tosf.96, 👆 bit.ly/tos16-3

K.H. Esbensen and C. Velis, "Transition to circular economy requires reliable statistical quantification and control of uncertainty and variability in waste. Editorial", *Waste Manage. Res.* **34(12),** 1197–1200 (2016). https://doi.org/10.1177/0734242X16680911, 👆 bit.ly/tos20-6

K.H. Esbensen, "Pierre Gy (1924–2015): a monumental scientific life", *TOS Forum* **Issue 6,** 39–46 (2016). https://doi.org/10.1255/tosf.92, 👆 bit.ly/tos20-7

21 Following the TOS will save you *a lot* of money (pun intended)

Partly by contrast with the previous chapter, below are topics that are *exclusively* positive and constructive. We have enlisted an imposing figure in the TOS panoply, Professor emeritus Pentti Minkkinen, erstwhile of Lappeenranta University of Technology, Finland, to take a first lead in this collaborative effort, due to his extensive experiences with the TOS. An initial survey of possible themes for this chapter quickly developed into a feast of possible title alternatives:

1) "You will not believe how much it costs not to follow the TOS"
2) "Following TOS will save you *a lot* of money (pun intended)"
3) "Save now—pay dearly later"
4) "It's not so expensive as you think to follow the TOS"
5) "The tighter the budget the more important is to use it wisely"

There can be no doubt about what is presented below then. "Follow the money" would appear to be a useful lead when matters of "the TOS or not?" come up. Below the reader is presented with clear cut case histories and examples all focusing on the potential for economic loss or gain—by following, or more importantly, by *not* following the TOS. Here are four business cases, specifically with in-depth TOS explanations—what more can one want?

AA
FOV: Field-of-View
IDE: Incorrect Delineation Error

Pentti Minkkinen

Technical TOS argument
Gather five sampling experts—and you will get at least six opinions as to the evergreen issue: "What works best to catch the attention of those who have never before been introduced to issues regarding, for example, sampling errors, sampling bias (and imprecision), the TOS, hidden costs, faulty decision making, economic risk?"

21.1 Case 1: Always mind analysis

Incorrect sampling operations can cause huge economic losses to industry. The impact from inferior, insufficient or incorrect sampling and assaying can be tremendous.

On the other hand, when the Theory of Sampling (TOS) is applied correctly, a considerable amount of money can be saved. Would you believe that a difference in average analytical values of only 0.06 % could accrue lost revenue of ~US$300 M in the mining industry? The mind boggles, but read on. The first case study focuses on the analytical part of the full sampling-and-analysis pathway in a copper mine operation.[1]

In 1970, a chemical laboratory of a large copper mining operation in Northern Chile was experiencing a bad response time problem because of the large amounts of samples to be assayed. The analytical method at the time was atomic absorption spectroscopy. In order to improve the performance, the chief chemist decided to change to X-ray fluorescence. The change reduced the cost and the response time. Only one geological matrix (high-grade secondary *sulfides*) was considered for calibration. Neither blind duplicates, nor standard reference materials were used at the time in order to monitor the precision and accuracy of the assaying process. This particular mining company was reasonably assured of its general performance because a considerable amount of effort and consulting had been spent to deal with the potentially fatal sampling issues if not complying with the principles laid down by the TOS. In fact this company rightly prided itself of this forward-looking attitude, which at the time was indeed quite extraordinary. All seemed to go the right way, but...

Meanwhile, exploration geologists were also beginning to send samples from a neighbouring deposit to the laboratory, for copper assay as well. This matrix was very different, however, copper *oxides*. Because this fact was *not* reported to the analytical chemist, the resulting assays, based on the sulfide calibration, turned out to be biased by 0.06% copper—when a later reckoning arrived. This bias may not seem to amount to much, but circumstances are deceiving.

Considering a yearly mining rate of 32 million tons, a recovery of 80%, an operational lifetime of 20 years, a price of US$1 per pound of copper (contemporary prices) and a discount rate of 10%, the economic bias caused by the analytical bias could be estimated to be US$ 292 M.

This was estimated as follows (the following equations are generic and can be used for quite a range of other projects in need of a similar economic evaluation):

$$B_i = [V_i(m) - p(t)] \cdot t_i \frac{1-e^{iN}}{} - I(t) \quad (1)$$

where B_i = net present value (M $); $V_i(m)$ = value of one ton of ore ($); $p(t)$ = cost of production of 1 ton of ore at $5 per ton; t = annual rate of production (ton/year) = 30 Mt/year; i = discount rate = 10%; N = life of mine (years) = 20 years; $I(t)$ = investments (M $) = $640 M; m = mean grade above cut-off grade (%Cu); $V_i(m) = 22.4 \cdot pr \cdot R \cdot m$; where pr = copper price = 0.8 $/lb, R = metallurgical recovery in % = $(m - 0.1008) \cdot 0.9/m$. From Equation 1, with a little rearranging:

$$B_1 = [V_1(m) - p(t)] \cdot t_i \frac{1-e^{iN}}{} - I(t) \quad (2)$$

$$B_2 = [V_2(m) - p(t)] \cdot t_i \frac{1-e^{iN}}{} - I(t) \quad (3)$$

$$\Delta B_i = \Delta V_i(m) \cdot t \frac{1-e^{iN}}{i} \quad (4)$$

$\Delta V(m) = 22 \cdot Pr \cdot R \cdot \Delta m$ then the economic bias loss in net percent values is, when inserted into Equation 4:

$$\Delta B_i = 1.056 \cdot 32 \cdot \frac{1-e^{-0.1 \cdot 20}}{0.1} \approx \$292 \text{ M}.$$

The lessons from Case Study 1 are very clear:

- The economic consequences of analytical biases can be of considerable magnitude. In the example of a low-grade mineral deposit, the magnitude was similar to the estimated profits!
- Analytical accuracy is essential for correct economic assessment of a mining project—but, of course, the ultimate target for accuracy is the target lot, the full mineralisation. Clearly the ultimate objective is to get a reliable assay in this context, not *just* a representative sample.
- Clear communication between corporate entities (sometimes organisational protagonists) is just as relevant as alignment with the overall business objectives.
- Systematic use of blind duplicate samples, reference materials (RM) and blanks is crucial in order to assure the quality of the *full* sampling-and-analytical process. This approach would have discovered the unfortunate consequences demonstrated in this case at a very early stage. For geologists as well as analytical chemists, it is inconceivable not to know the dangers of analysing oxides *as if they were* sulfides. It was the simple lack of inter-departmental communication that led to this dramatic economic loss.

21.2 Case 2: Saving a client from a wrong, expensive investment

Finely ground limestone is much used as a high-quality *coating* in the paper industry. But accidental "coarse particles" (i.e. particles larger than 5 µm) in coatings often result in severe defects in high-speed printing machines. These may actually break the paper web, which leads to very expensive stops in production that must be avoided literally "at all costs". There is no room for neglecting this problem in the paper printing industry. The quality of the coating product must comply with the stringent demand of less than five such particles in every ton of ground limestone coating.

The manager at a limestone coating producer was considering buying an expensive particle size analyser for on-line quality control in order to deal effectively with this issue. The question was would this be an economically viable solution?

Let's ask the TOS. Part of the armament of a sampling expert is a thorough knowledge of the features and the use of the classical statistical Poisson distribution. For the uninitiated, here is a situation in which *Wikipedia* is just the right source for competent, compact information:

*In probability theory and statistics, the Poisson distribution, named after French mathematician Siméon Denis Poisson, is a discrete probability distribution that expresses the probability of a given number of events occurring in a fixed interval of time, or space, if these events occur with a known, constant rate and **independently** of the time since the last event. The Poisson distribution can also be used for the number of events in other specified intervals such as distance, area or volume. For instance, an individual keeping track of the amount of mail received each day may notice that he or she receives an average number of four letters per day. If receiving any particular piece of mail does not affect the arrival times of future pieces of mail, i.e., if pieces of mail from a wide range of sources arrive independently of one another, then a reasonable assumption is that the number of pieces of mail received in a day follows a Poisson distribution. Other examples that may follow a Poisson include the number of phone calls received by a call center per hour or the number of decay events per second from a radioactive source. (Wikipedia, accessed 23 March 2018).*

And, in the present case, the number of adverse coarse particles found in a volume of coating material is also likely to follow a Poisson distribution. This gives the sampling expert just the right weapon with which to offer to evaluate the economics of the suggested acquisition of the expensive on-line particle analyser.

First, as in every situation in science and industry, the problem should be as clearly defined as possible. In this case the target question can be stated either 1) as the acceptable relative standard deviation of the measurement, or 2) as the risk of confidence that the target value of five particles/ton is *not* exceeded. This translates into two classical questions:

1) How big a sample is needed if a relative standard deviation of 20% is acceptable?

It is easy to apply the Poisson distribution, because one of its well-known features is that the relative standard deviation (s_r) is inversely proportional to the square root of the number of observed events (n):

$$s_r = \frac{1}{\sqrt{n}} = 20\% = 0.2$$

From this we obtain that the sample should be so big that it contains $n = (1/s_r^2) = 1/0.2^2 = 25$ particles. At the critical concentration of five particles/ton, this means that the sample size needed is five tons! This magnitude was not known to the producer before inviting a sampling expert to evaluate the case. Obviously, this first result made him flinch.

Alternatively, the question can also be stated:

2) What is the highest number of large particles in a 1-ton sample that can be accepted in order to be able to guarantee that the product is acceptable, if a 5% risk of a wrong decision is acceptable?

From the Poisson distribution we obtain (with a little probability calculus thrown in, not included here) that the probability for one particle (or less) is 4% and for two particles (or less) it is 12%.

So, even a reliable sample, a representative sample of 1 ton will *not* be able to satisfy the very strict demands here. Besides how to ascertain, how to measure the number of coarse particles in such a huge sample? Extreme practical sieving would probably be the only rigorous way—not exactly what was hoped for with an on-line particle size analyser!

In detail, this example shows that even if you buy the most expensive state-of-the-art particle analyser, it would be completely useless—*just money wasted*. Here is the reason why: particle size analysers are designed to handle samples of the size of ~ a *few grams* only! Neither is there a sieve system that could separate out just a few 5 μm size particles from tons of fine powder (here one needs to invoke a little standard powder technology knowledge, which is readily available for the inquisitive consultant).

21.2.1   Conclusion

The only way to maintain product quality is to do regular checks and maintenance of the actual production machinery. The only way to study ton-sized samples of this kind of powder material is to build a *pilot plant* where large-scale coating experiments can be run effectively. While this is indeed an expensive solution for the coating producer, the customer has the same problem. He cannot complain about the quality of the received material based on his own in-house analytical measurements for the exact same reasons. Of course, the customer may well *suspect* the coating material quality, if there are too many breaks in the paper web after the coating is applied or if the paper maker gets complaints from the printer using his paper, but nothing can be *proven* with statistical and scientific certainty because of the limitations revealed by thorough application of the TOS and classical statistics. The only way forward is to suggest that the producer and the customer *together* engage in establishing a joint pilot plant for practical systematic testing. It *may* perhaps be possible to do some fancy down-scaling in this context, but this is another story altogether.

21.3 Case 3: The hidden costs—profit gained by using the TOS

An undisclosed pulp mill was feeding a paper mill through a pipeline pumping the pulp at about 2% "consistency" (industry term for "solids content"). The total mass of the delivered pulp was estimated based on the measurement of a process analyser installed in the pipeline immediately after the slurry pump at the pulp factory. Material balance calculations showed that the paper mill could not produce the expected tonnage of paper based on the consistency measurements of the process analyser.

When things became too difficult to proceed, an expert panel was convened to check and evaluate the measurement system. A careful audit complemented with TOS-compatible experiments revealed that the consistency measurements were *biased*, in fact giving 10% too high results. The bias originated from two main sources:

1) The process analyser was placed in the wrong location and suffered from a serious *sample delimitation error* (often the weakness of process analysers installed in pipelines); the on-line analyser field-of-view (FOV) did *not* comply with the TOS' fundamental stipulation regarding process sampling, that of corresponding to a full slice of the flowing matter, thus a substantial IDE was created (see Chapter 16 or Reference 2).

2) The other error source concerned the process analyser calibration. It turned out that the calibration *depended* on the pulp quality (never an issue it was thought of at the time of the original calibration—but softwood and hardwood pulps need *different calibrations*).

By making the sampling system TOS-compatible and by updating the analyser calibration models (competent chemometricians were called in), it was possible to fully eliminate the 10% bias detected.

Payback: It is interesting to consider the payback time for involving the service of the TOS and chemometrics in this case. The pulp production rate was about 12 ton h^{-1}. The contemporary price of pulp could be set as an average of $700/ton, so the value produced per hour was easily calculated to be $8400 h^{-1}. The value of the 10% bias is thus $840 h^{-1}. As the cost of the evaluation study was about $10,000 the payback time of this investigation was about 12 h. To be strict, the costs for the TOS-compliant upgrading of the process analyser should be added, but in this case corresponded to only a few weeks of production (and it should rightly have been covered by the original installation costs).

It does not have to be expensive to invoke proper TOS competency—not at all!

21.4 Case 4: The cost of assuming standard normality for serial data

Data from natural processes and especially manufacturing industrial processes are by their nature very nearly always *auto-correlated*. This is an intrinsic data feature that can be used to great advantage—but this is a great *disadvantage* if not understood and used properly. Thus, there is a widespread *assumption* that the variation of this kind of serial data can be well approximated by straight forward application of the standard normal distribution. This is a distinctly dangerous assumption, however, that will always lead to sub-optimal (and expensive) sampling plans for estimation of average lot values, at a(ny) given uncertainty level. This last case illustrates the consequences of accepting this persistent, but wrong assumption.

The data used here are recorded from a wastewater treatment plant discharge point; measurements were carried out once per day. This data series is used in order to estimate the amount of sulphur that is discharged annually into a recipient lake.

For this purpose, a standard variographic experiment was carried out collecting and analysing one sample per day—for 30 days. Figure 21.1 shows the process data, expressed as *heterogeneity contributions*, and the variogram calculated based hereupon. The heterogeneity contribution of a measurement (h_i) is defined as the relative deviation from the mean value (a_L) of the data series:

$$h_i = \frac{a_i - a_L}{a_L}$$

(see Reference 3 or earlier chapters in this book).

The most important aspect of the TOS' variographic characterisation facility is the realisation that the uncertainty of the mean values of auto-correlated series *depends* on the sampling mode, which can be *random (ra)*, *stratified random (str)* or *systematic (sys)* sampling. By analysing the variogram, variance estimates are obtained which subsequently are used to calculate the variance of the process mean:

$$s^2_{a_L} = \frac{s^2_{mode}}{n}.$$

Figure 21.2 presents the relative standard deviations as a function of the unit measurement sampling lag (multiples of 1 day) for systematic and stratified random sampling.

If the normality assumption is used, then the relative variance estimate is the variance estimate calculated from the 30 measurements without any consideration as to their order (their auto-correlation). The relative process variance calculated on this basis is 0.0820, which corresponds to a relative standard deviation $s_r = 28.6\%$.

If, after this experiment, a sampling plan is made with the purpose of estimating the annual discharge using, for example, one sample per week, the relative standard deviation estimates can

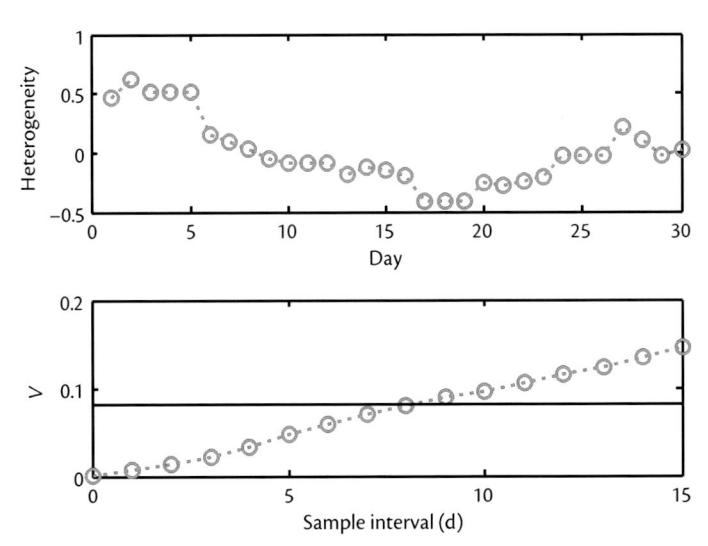

Following the TOS will save you a lot of money (pun intended)   257

Figure 21.1. Variation of the sulphur content of a wastewater discharge, expressed as heterogeneity contributions over 30 days (upper panel) and the corresponding variogram (lower panel). The black line in the lower panel represents the overall process variance, i.e. the variance of all 30 heterogeneity values as treated by standard statistics with no auto-correlation considerations.

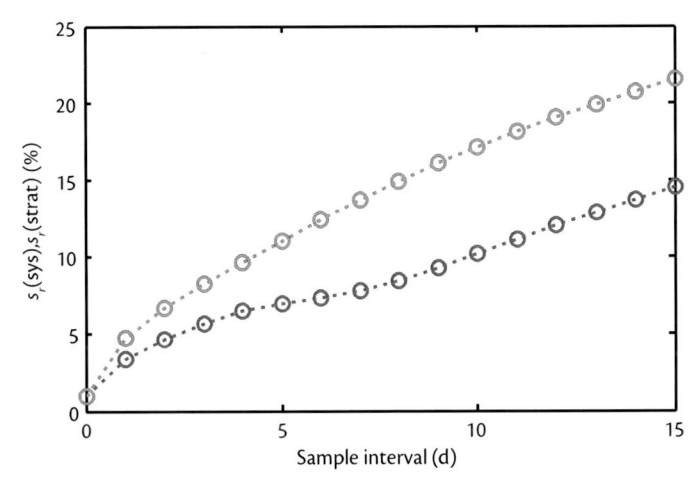

Figure 21.2. Estimates of the relative measurement standard deviations for systematic (darker grey) and stratified random (lighter grey) sampling as function of the sample lag (one-day intervals).

be obtained from the variogram at sample lag 7; they are 7.8 % for systematic sample selection and 13.7 % for the stratified random mode.

If, instead of depending on a daily sample, say, it is decided to collect just one sample per week for the monitoring purpose, then the number of samples in estimating the annual average will be $n = 52$. Now the following results are obtained:

- Relative standard deviation of the annual mean by *systematic sampling*:

$$s_{a_L} = \frac{s_{sys}}{\sqrt{n}} = \frac{7.8\ \%}{\sqrt{52}} = 1.1\ \%$$

- The expanded uncertainty of the annual mean from the systematic sampling is $U = 2 \cdot s_{aL} = 2.2\ \%$
- Relative standard deviation of the annual mean of *stratified sampling*:

$$s_{a_L} = \frac{s_{strat}}{\sqrt{n}} = \frac{13.7\ \%}{\sqrt{52}} = 1.9\ \%$$

- The expanded uncertainty of the annual mean from the stratified sampling is $U = 2 \cdot s_{aL} = 3.8\ \%$

Clearly, the *systematic sampling mode* in this case will be the method of choice.

Many current sampling guides still advise that the required number of samples for a targeted uncertainty is estimated by using the standard normal distribution approximation based on all available data *without* taking auto-correlation into account. Then the relative standard deviation of the process data will be $s_r = 28.6\ \%$. If this is used the following will result, if based on the same uncertainty level as for the systematic sample selection:

$$n = \frac{s_r^2}{(1.1\ \%)^2} = \frac{(28.6\ \%)^2}{(1.1\ \%)^2} = 678$$

which is a very different number of "required samples" than 52!

21.4.1 Conclusion

The standard normal distribution assumption is *expensive*—to say the least! The time used to collect and analyse over 600 *extra samples* could most certainly be used in a much more profitable way. Knowing well the distinction between a random set of data (for which classical statistics is the correct tool) and a *serial set* of process data (or similar—it is the auto-correlation that matters) is another element in the tool kit of the competent sampler (process sampler in this case). Variographic characterisation is a very powerful part of the TOS.

21.5 Lessons learned

- Incorrect sampling and ill-informed analysis generates *hidden losses* that do not appear in accounting spreadsheets, for which reason top management do not easily become aware of them.
- There is a "natural tendency" to focus on *effects*, and **not** on *causes* of problems. This attitude creates unhappiness, time and economic losses, while not solving anything.
- If one does not fully understand all *sources* of variability of industrial processes etc., losses are difficult to discover and their economic impacts are difficult to estimate. Yet this quantification is precisely (very often the *only*) manifestation that top management wishes to see.
- When there is little communication between different professions, there is deep trouble. Many professionals are primarily focused on solving their own problems, which are not necessarily always aligned with the objectives of the company, which is to produce high-quality products at the lowest possible cost. As a consequence, many professionals do not collaborate well with each other, they do not know each other and each other's problems well, or at all.

So what works best?
Still difficult to say—but we will soon get a chance to appraise the reactions from the first generation of readers of this book ... Stay tuned, e.g. in *Spectroscopy Europe* (https://www.spectroscopyeurope.com/sampling, 👆 bit.ly/tos20-2) or *TOS Forum* (https://www.impopen.com/tos-forum, 👆 bit.ly/tos21-a).

- Up to the turn of the millennium, a largely isolated sampling community was unable to communicate effectively the relevance of sampling to top management in economic terms. But, for individual consultants, the field was wide open. It was a very good time for them!
- Initiation to the TOS has always been considered "difficult" in many sectors in science, technology and industry, because the fundamental texts were often felt to be cryptic, or (quite) a bit too mathematical (for most), and hence not easy for beginners to understand.
- However, there has been a revolution is this context in the years since the first World Conference on Sampling and Blending (2003), from which a seminal contribution is particularly relevant.[4] Today, there is an abundance of easy introductions and texts to be found, the quality of which is excellent. It is hoped that the present introductory book will contribute to drive this development even further.

21.6 References

1. P. Carasco, P. Carasco and E. Jara, "The economic impact of correct sampling and analysis practices in the copper mining industry", in "Special Issue: 50 years of Pierre Gy's Theory of Sampling. Proceedings: First World Conference on Sampling and Blending (WCSB1)", Ed by K.H. Esbensen and P. Minkkinen, *Chemometr. Intell. Lab. Sys.* **74(1)**, 209–213 (2004). https://doi.org/10.1016/j.chemolab.2004.04.013, 👆 bit.ly/tos21-1
2. K.H. Esbensen and C. Wagner, "Process sampling: the importance of correct increment extraction", *Spectrosc. Europe* **29(3)**, 17–20 (2017). 👆 bit.ly/tos21-2
3. P.M. Gy, *Sampling of Heterogeneous and Dynamic Material Systems*. Elsevier, Amsterdam (1992).

4. P. Minkkinen, "Practical applications of sampling theory", in "Special Issue: 50 years of Pierre Gy's Theory of Sampling. Proceedings: First World Conference on Sampling and Blending (WCSB1)", Ed by K.H. Esbensen and P. Minkkinen, *Chemometr. Intell. Lab. Sys.* **74(1)**, 85–94 (2004). https://doi.org/10.1016/j.chemolab.2004.03.013, ☞ bit.ly/tos21-4

22 A tale of two laboratories I: the challenge

This is a tale of two fictional commercial laboratories, but all features represent true events and occurrences taken from a range of real-world laboratories, only here re-arranged in a more focused fashion for a purpose: "How can the Theory of Sampling (TOS) help the commercial laboratory to improve its reputation and to increase its business"? The relevance for existing laboratories is striking. The reader will have to bear with this chapter for mostly focusing on sampling issues in what obviously is a much more complex scientific and business context, but please indulge us for a little while—there is a sharp return to real-world, analytical realities at the end of the story.

Laboratory A is in fierce market competition with Laboratory B (and indeed several others in the global market), a situation that has existed for decades. This has so far led to a healthy business-oriented science, technology and human capital drive that has served both laboratories well. Both laboratories are also keenly aware of the necessity to be in command of the TOS for all relevant in-house activities involving sampling, sub-sampling, mass-reduction, sample preparation etc. But whereas Laboratory A has availed itself of the services of the TOS strictly within its regimen only (as is indeed the case for most laboratories), one fine day the manager of Laboratory B had an epiphany that made her see the potential advantages of applying the TOS in full, which involved a distinctly "beyond-the-laboratory" scope. What

AA

GP: Governing Principles
SUO: Sampling Unit Operations
MU: Measurement Uncertainty
TSE: Total Sampling Error
TAE: Total Analytical Error

If the reader heeded the initial suggestion of reading Chapters 22–24 between Chapters 2 and 3, there is no justification for not reading them again. The many critical TOS issues in the "Tale" and its resolution will now be totally clear and even more compelling in context.

happened on that fine day? And how did it help Laboratory B to do better in the market?

> Disclaimer: There is no identifiable, real-world laboratory that corresponds completely to Laboratory A; it is for the tale's convenience that individual features met with in several real-world laboratories have been aggregated under this generic name. Thus, Laboratory A mostly exists as a collection of issues, **some** of which may happen to also characterise the reader's laboratory, while Laboratory B is of an altogether different nature. Laboratory B is an emerging entity, hopefully on the verge to becoming real. It could be yours—or your company, organisation or institution could become the first mover!

22.1   Introduction (scientific, technological)

The traditional view of the role, tasks and responsibilities of the analytical laboratory is well described by Laitinen in an Editorial in *Analytical Chemistry* in 1979.[1] Despite being written 40 years ago, this is still an apt summary of the workings, and frustrations, of *laboratory life*. The emphasis on lack of communication between stakeholders is still very much true today. The real flag-raiser is hidden in the following sentence: *"...the analyst chemist is frustrated... not being informed about the full background of the sample, its urgency, or the use to which the measurements will be put"*. Based upon the body of knowledge presented in the previous chapters in this book, it is obvious that the analytical laboratory is a very important area for application of the TOS. There is so much *essential* sub-sampling going on, and there are many aspects of sample preparation and presentation that involve elements of mass-reduction as well (which is also sampling). In short, the analytical laboratory is also on the frontline regarding the application of the TOS.

Read the Laitinen editorial in full at https://doi.org/10.1021/ac50047a600, bit.ly/tos22-a

Of course, analytical chemistry reigns supreme in the laboratory. Full, always updated, command of the science and technology of analysis (analytical chemistry, physical characterisation, other) is the *raison d'etre* for all analytical laboratories, research or commercial. However, proper sampling also plays a critical role, as has been made abundantly clear in the present book regarding *why?* and *how?* to sample. It is indeed fully possible to make a mess of the internal mass-reduction pathway towards analysis with severe breaks with representativity **if** the elements of the ongoing sub-sampling are not in compliance with the stipulations in the TOS. The problem has been, and still is, that such operations are traditionally viewed as but "trivial" mechanical parts of the analytical process, and consequently have only rarely, if at all, been viewed in a systematic fashion—enter the TOS. The present tale revolves around these issues.[2]

22.2 There is analysis… and there is analysis+

As soon as *samples* (primary samples, or suitably sub-sampled proportions thereof) have been delivered to the in-door of an analytical laboratory, sub-sampling is in fact **the** critical success factor that must be dealt with in a proper fashion in order to be able to document relevant, reliable, representative analytical results.[3,4] This is a subtlety often not fully acknowledged, or even recognised, in the flurry of business and economic optimisation taking the driver's seat in the laboratory. It is all too often *assumed* that all operations and facilities needed for proper analysis are fully known and tested in the operations of a professional laboratory: procedures, equipment, work-paths, training—for which reason laboratory efficiency is typically only considered within the narrow scope of organising and optimising the complex analytical workflow. Truth be told, this is far from a simple matter in practice, but this is exactly where good managers have the opportunity to shine. However, in this managerial view, the elements in all analytical workflows are **fixed** and fully optimised, so that sample *throughput* comes into focus as the key factor of interest. Such elements are sample receiving, initial sample preparation, sub-sampling, possible sending off of sub-samples to other within-company laboratories for other types of analysis, more sample preparation (this time for analytical support), e.g. grinding, milling, mixing, sieving, analyte extraction (for some types of analysis only) and many other specialised part-operations where/when needed. The key point here is that all these elements are considered as objects that can, indeed *should*, be managed *exclusively* according to a flow-path business objective with the aim of producing the necessary economic profit. There are certainly also other issues in play here, but these are necessarily also usually considered from the economic profitability point of view.

Call this position "Laboratory A" in this tale. It is likely a fair statement that this is the main business strategy behind most of today's commercial analytical laboratories. But under normal market conditions profit will not sky-rocket for Laboratory A as a function of even the best manager's efforts, because of the relentless strong market competition. For the time being, this is the same basis upon which Laboratory B is in business as well.

Thus, at the outset, Laboratories A and B are competing on what economists call a fair basis; both have competent managers and highly competent technical staff (scientists, technicians, managers…), and superior logistics. They both also have access to the same external (public domain) scientific and technological facilities and developments with which to sustain their current endeavours and, hopefully, improve their state-of-the-art capabilities, and thereby be able to drive their individual businesses forward. So how is it possible for an individual laboratory to increase its market share? Better marketing and better presence in the market are the first business-related options with which to promote its potentially better offerings. The archetypal answer in a high-tech context is either by becoming more *efficient* than competing companies and/or through being the *first mover* with respect to significant technological developments, perhaps even disruptive technological breakthroughs. Both laboratories will, in principle, have equal access to all developments which are published as a result of university R&D, but it **is** possible that the ability to spot what may *become* new disrupting developments could be different in the two laboratories. Also, much of analytical technological development takes place **within** the laboratory regimens, and here it is everybody's game. However, the present tale is not about possible new analytical technological developments or potential comparative advantages. The core element in this tale has been around for more than 60 years, and could thus not possibly contend with modern, disruptive technology breakthroughs—or could it?

22.3   The core issue

The core issue is what turned out to develop into a marked *difference*, which only Laboratory B decided to take advantage of. At its root, it is based on the in-depth understanding ("technical" understanding, if you will) that stems from the TOS:

"The quality and relevance of an analytical result is not **only** a function of the analytical competence, analytical equipment, work-path optimisation, i.e. the traditional business understanding which is the firm position of Laboratory A. It is **equally as much a function of the specific sampling procedures** a.o. involved in producing the analytical aliquot."[2–5]

Thus, while precision is a quality characteristic of the analytical method, and while the analytical accuracy refers to the mass that has actually been analysed **only**, i.e. to the mass of the aliquot, this is not in compliance with the *needs* of the users of analytical results, the clients of the laboratory. Users will invariably make important decisions (sometimes trivial decisions, at other times truly critical decisions, sometimes even life-and-death decisions in science, technology, industry and society) based on the **reliability** of the analytical results. In an overwhelming proportion of cases this is tantamount to the *reputation* of the analytical laboratory. Indeed, the specific analytical quality and performance characteristics are the central focus for all kinds of official or commercial auditing, accreditation and certification of laboratories (CEN, ISO, JORC etc.). The present tale focuses on the fact that all of the above demands are declared as satisfactory as long as the *analytical* accuracy and precision meet relevant criteria.

> But the core issue of this tale constitutes a different scope. The practical focus of the user is the *original lot*, **not** the analytical aliquot. Enter the critical factor of *how* the enormous gulf between lot and aliquot has been bridged. This has everything to do with the ability to comply with the TOS in all stages of the pathway from-lot-to-aliquot, Figure 22.1 and 22.2.

22.4 The crux of the matter

The user who is to make critical decisions does not care one iota how accurate a specific analytical method is with respect to the *miniscule* analytical mass! The decision maker is **only** concerned with how accurate a particular analytical result is (say, 3.57 % or 276 ppm) *with respect to the original lot*. The operative question for all users is: "how accurate is this compositional determination with respect to the original lot?". How can I be certain this is a determination that holds up with respect to the 1000-to-100,000 times *larger* original lot? What is the uncertainty related to the

A tale of two laboratories I: the challenge

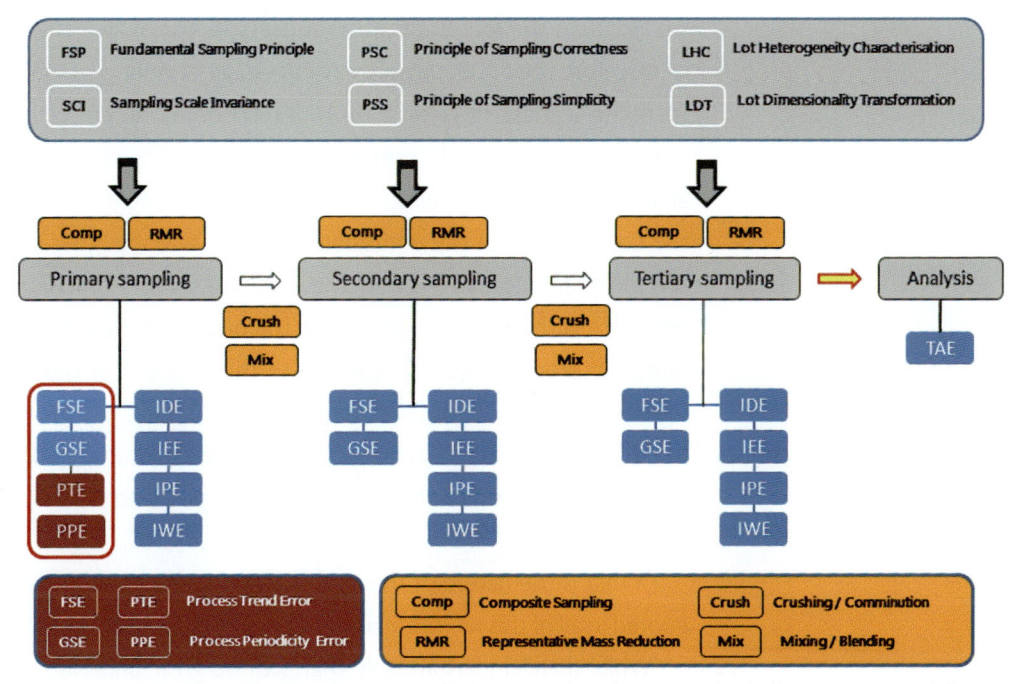

Figure 22.1. A comprehensive summary of the TOS, the elements of which are comprised by six Governing Principles (GP, grey), four Sampling Unit Operations (SUP, yellow) and six (eight) sampling errors (blue, maroon) and their relationships with respect to the multi-staged sampling process "from-lot-to-aliquot". See TOS references[2-9] for in-depth description. Illustration copyright KHE Consulting; reproduced with permission.

analytical result **in this context**? This focus is very real and leads to the above questions, which need definite answers.

But the way clients and laboratories traditionally go about this issue lacks relevance and rigour. The tradition is to point to, and rely on, official laboratory accreditations, analytical performance evaluations etc. But this kind of validation, verification and justification is exclusively based on the narrow analytical accuracy and precision characteristics—which all focus on the **aliquot**.

However, the analytical aliquot mass/volume is very far from the legitimate practical concern, the lot. The analytical aliquot is typically (on a mass/mass basis) some 10^3, 10^6 (or more) times *smaller* than the lot. The **critical success factor** of all analysis is, therefore, that the complete, multi-stage sampling process spanning a mass reduction of 1/100,000 or more, *preceding* analysis, is scrupulously **representative**. How else could the analytical result of the aliquot say anything meaningful about the composition of the entire lot? The only available **guarantee** for representativity (simultaneous accuracy and precision relevance with respect to the lot) is the quality of the specific sampling process used to cover these three to six (or more) orders-of-magnitude of mass reduction before analysis. From the TOS it is known that it is only a specific, documentable *sampling process* that can be evaluated, assessed and declared to be representative, or not.[a]

"What is the nature of the accuracy and precision estimates quoted in all of this world's analytical laboratory accreditations?"[3]

While there understandably may well be great pride in the analytical capabilities of a(ny) specific analytical laboratory… the *relevant* decision-making issue, relevant for the user of analytical results, is very often *missing* from current analytical reports and certifications.[3] Typically an estimate of the operative, real-world decision-making accuracy *with respect to the original lot* is nowhere to be found, very likely because covering this would entail that the laboratory should be involved with all sampling going on, specifically also that associated with the primary sampling from the lot.[3-5] This would require the laboratory to be(come) fully TOS competent. A graphic summary of the boundary conditions for this challenge is given in Figures 22.1 and 22.2.

---

[a] It is not possible to subject "representativeness" to grammatical declination—a sampling process either **is** representative, or it is **not**…

Adverse charateristic: Heterogeneity

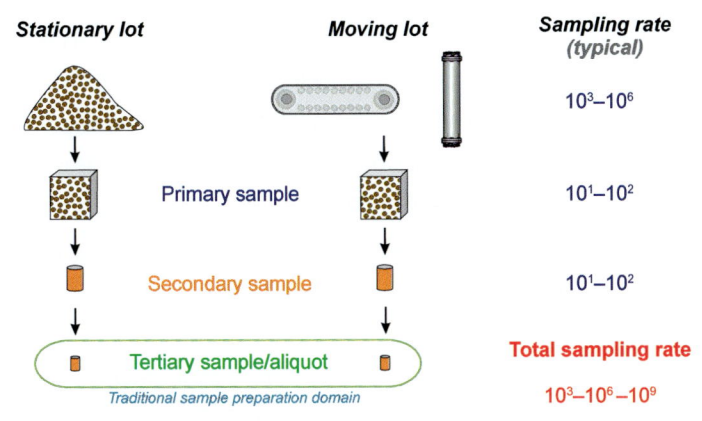

		Sampling rate (typical)
Stationary lot / Moving lot		10^3–10^6
Primary sample		10^1–10^2
Secondary sample		10^1–10^2
Tertiary sample/aliquot *(Traditional sample preparation domain)*		**Total sampling rate** 10^3–10^6–10^9

Figure 22.2. The general from-lot-to-aliquot pathway encompasses sampling processes which are in no way simple mass-reductions only, but which require complete compliance with the TOS at **all** sampling stages, as summarised in Figure 23.1.[2-5]

22.5 The complete argument

There are always several *possible* different sampling methods that can be used in a particular situation, at a particular scale—first and foremost grab sampling vs composite sampling, or composite sampling based on a significantly different number of increments vis-à-vis the lot heterogeneity addressed. In the case of, typically, three stages of sampling and sub-sampling in the laboratory, there are many possibilities for coming up with functionally different sampling pathways from-lot-to-aliquot. All will lead to an analytical aliquot, but the analytical results will *per force* be different, which specifically is a matter of the degree of successful heterogeneity countermeasures embedded in the sub-sampling at particular sampling stages. Thus, in a very real sense the specific sampling pathway will influence the analytical results—an aliquot is not just an aliquot that can be considered in isolation, all aliquots have a past, a provenance.

> Thus, a fundamental tenet stemming from the TOS is that all analytical results are but *estimates* of the composition of the original lot. Hopefully the best possible estimate of course, but "best" is not an automatic qualifier. "Best" specifically means, and should only mean, that the analytical report reflects the singular representative analytical result that directly can be used for the important societal, corporate, environmental decision making. Which again brings forth the key understanding that the qualifier "representative" is related to the perspective of the *complete* sampling process "from-lot-to-aliquot-to-analysis", **not** to the infinitely smaller context "from-aliquot-to analysis" only. It matters, crucially, *how* the analytical aliquot was arrived at.[2-9] All laboratories must be sufficiently TOS competent.

In fact, it is fully possible to make use of *bona fide* analytical methods (likely with extremely good analytical accuracy and precision), which in the absence of a preceding representative sampling process, may easily end up as having the quirky characteristic of delivering analytical results that are "precisely wrong". This surprising understanding concerns the fundamentally different nature of the analytical vs the sampling + analytical bias, an issue which has featured extensively in recent TOS literature.[3-5] Figure 22.3 depicts this crucial distinction graphically. Competent understanding of the nature of the sampling bias is an absolute necessity for anyone claiming to be a sampler. The reader is encouraged to study Figure 22.3 and its caption very carefully.

22.6 The meaning of it all

From the TOS' five-plus decades of empirical experience, see for example the sampling standard DS 3077,[4] it is well known that the uncertainty stemming from a *preceding* sampling stage *on*

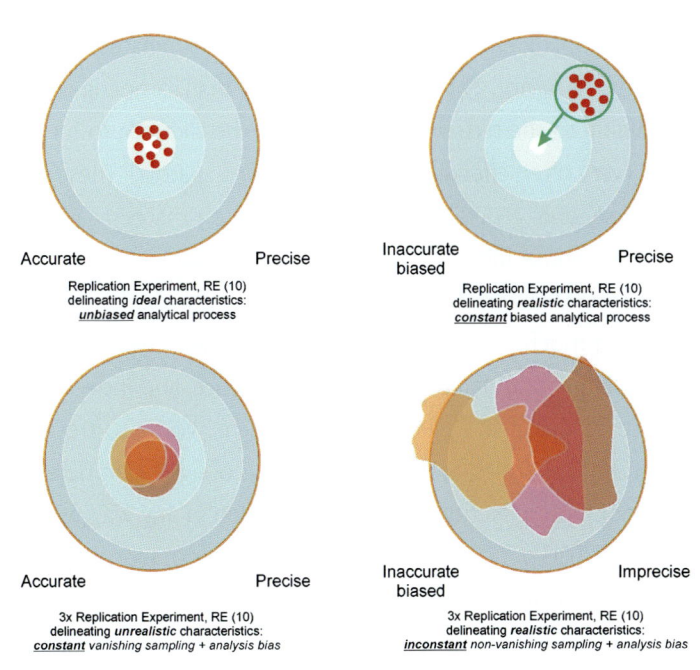

Figure 22.3. While an analytical bias can always be subject to a statistical bias-correction (upper panels), the nature of the sampling + analytical bias is fundamentally different (lower panels). Because of the interaction between a specific material heterogeneity and a specific sampling process, which may be more-or-less removed from the qualifier "representativity", replicated sampling + analysis will always result in a different accuracy and precision estimate; thus the sampling + analysis bias is *inconstant*.[3] Illustration copyright KHE Consulting; reproduced with permission.

average can be up to 10× **larger**.[b] Thus, if not TOS compliant, the sampling error uncertainty stemming from the primary sampling stage operations is *on average* 10× **larger** than those originating at the secondary sampling stage. Indeed, it may be even larger, depending on the material heterogeneity and the degree with

---

[b]The relationships between sampling—and analytical *errors* and their effects on Measurement Uncertainty (MU) is treated in a benchmark paper in great detail; interested readers are referred to Reference 3.

which sampling errors have been adequately eliminated and/or suppressed, or not. This again is 10× **larger** than sampling errors pertaining to the tertiary sampling stage, the aliquot-producing stage (very often effected by a grab sampling spatula). These are, of course, only general order-of-magnitude estimates. Materials will exist whose inherent heterogeneities would lead to somewhat *smaller* factors, but there just as assuredly also be materials with a much more troublesome heterogeneity, which would lead to larger-than factors of 10×. Best practice, and indeed the only strict logical and scientific attitude here, will be to treat all materials, at all scales, as if they were *always* of significant heterogeneity. Upon reflection, this is not a burden in any way, as the pertinent TOS principles, unit operations and error management rules needed are scale invariant.

What appears to be the saving grace is that all the world's laboratories can safely be assumed to have *minimised* their within-house analytical uncertainties to the greatest possible extent which, alas, is but a small fraction of the TSE. Thus, the core *message* from the TOS' experience is that there is a step-up, potentially up to several orders-of-magnitude, of the sampling uncertainty accumulated over all effective sampling stages preceding analysis (the total sampling error, TSE), in the **absence** of any specific heterogeneity counteraction. This counteraction is the *raison d'etre* for the TOS. Either way, TSE always *dominates* TAE, occasionally to such a degree that TAE dwindles into insignificance. The point is that a situation very rarely exists in which TAE is even close to TSE in magnitude. When such is the case, this would signify a laboratory truly very much in the lead, because all in-house sampling errors would have been completely minimised.

But how would a laboratory go about proving this? A survey of reputable analytical laboratory websites is very telling, mostly because of a certain sin-of-omission regarding estimates of the effective TSE accumulated over all sampling stages. Instead there is often a disclaimer to be found in analytical certificates, that the analytical results only pertain to the aliquot (or only to the samples received)—which really only shifts the burden away from the laboratory back to the client, who may, or may not, be aware of this. The responsibility for procuring valid and reliable, representative analytical results has thus conveniently been shifted into a no-man's land. But who will take up the gauntlet of getting the client out of this predicament?

22.7 Inside and outside the complacent four walls of the analytical laboratory

Thus, the real culprit, the core issue of this tale, would **still** not have been addressed, because this lies *outside* the traditional laboratory regimen. The somewhat uncomfortable summary effect of all of the above is that the *primary sampling* stage very nearly always dominates TSE all by itself. But this is almost never included in laboratory performance reports.

For two reasons:

i) Due to market competition, the responsible analytical laboratory will always tend to have the smallest possible residual uncertainty from all the operative steps involved in its many different analytical offerings to its clients. For *fully* responsible laboratories this includes a genuine focus on minimising the tertiary and often also secondary in-house sampling (in reality *sub-sampling* in relation to the not-yet-included primary sampling from the lot). This aspect is what will differentiate between individual laboratories, which *may* decide differently as to what degree to also venture outside the laboratory when **full** TOS optimisation is acknowledged.

ii) The second reason is that Laboratory A deliberately declares: "Primary sampling is **outside** our responsibility". This is the hidden elephant in the room. Analytical laboratories may, or may not, deliberately consider that all ex-laboratory issues per definition are irrelevant—while the reality for *users* of the analytical results are completely dominated by the contribution from this "missing link". This issue is actually the **only** discriminating issue between the generic Laboratory A and Laboratory B.

What happened to generate this potential difference?

22.8 "One fine day"...

"One fine day" the manager of Laboratory B called in at work consumed with a completely new attitude, based on an epiphany she had had in her dreams the night before. Barely in the door, calling an immediate board and section chiefs meeting, the manager declared (eyes shining with newfound *righteousness*):

"There is a completely unrecognised business opportunity that no other laboratory has tapped into... yet. Laboratory B must be the first mover, Laboratory B must be the first to reap this competitive advantage! It has dawned upon me that despite Laboratory B's most stringent efforts to curtail all total in-house errors, we have erred, believing that this was well summarised by TAE... We have erred grossly! It is in reality [TSE + TAE] that is accountable for all the real-world's 'analytical variance'. It has dawned upon me that we are at least a factor 10× too low in our declarations in our analytical certificates—and depending upon the heterogeneity of lot materials and the ability to follow the TOS, this factor could be higher!" (the manager shuddered visibly). "What's more—today we have absolutely no, or only very little, possibility to influence this issue since this problem originates with/at the primary sampling from the original lot, which this laboratory so far has declared to be exclusively the responsibility of the client. How often does our Laboratory B, which we like to call 'the leading laboratory in the world', insist that it is in fact also *our responsibility* to explain to clients that this is a problem of significantly larger impact with respect to the interpretability of the analytical results *in context* than any other? A factor of 10+ or larger."

The above account, actually not a tale **at all**, has gone to great efforts to explain the "technical" TOS-based evidence for the situation revealed: as long as very many still do not take primary sampling sufficiently seriously (neither clients, nor laboratories), this should rightly be called the primary sampling disaster! As long as this has not happened, what are the consequences?

They are numerous, and they concern both company bottom lines and laboratory efficiency, which are all direct economic negative outcomes. They also concern the possibilities for necessary and efficient societal and public regulation and control (e.g. food, feed, pharmaceutical drugs, public health etc.), and here with likely much larger negative economic impacts, although often *hidden* at first sight. And they concern the reputation of the analytical science, technology and trade—which in the end reflects on the reputation of each individual analytical laboratory (commercial or not).

22.9 The really important aspect: costs or gains

Could there really be direct economic and business advantages in taking on the primary sampling issue—an issue so long considered as "**not** within our laboratory's responsibility"? The most often heard "justification" used in this context (remember that every single feature in this compound tale is **true**) is:

"This laboratory need not concern itself with primary sampling... This will cost us additional work, man hours, expenditures. This will break up our established work paths—all of which will impact negatively on our bottom line. And we will especially not be involved in this matter, since none of our competitors take this up either—we would simply be losing money in-house, and to no business advantage!"

The world's laboratories, clan A, have spoken!

This is the *status quo* in a large segment of the commercial analytical laboratory realm.

Nevertheless, Laboratory B decided to be the *first mover* and to proceed down this new road.

22.10 What in the world?

What was the *epiphany* experienced by the manager of what became: Laboratory A → Laboratory B?

What was the *business argument* that negated the above justifications for doing nothing, for continuing exactly as before—for continuing exactly as all the other, competing laboratories?

What will it take to seize the day?

Will your laboratory become Laboratory B tomorrow?

The following proverb is attributed to the founder of the TOS, Pierre Gy. Think of this in relation to the dominating primary sampling error/uncertainty! "SAMPLING—is not gambling!" Pierre Gy (1924–2015).

One may also factor in a well-known contradiction regarding human capital management:

"CFO asks CEO: 'What happens if we invest in developing our people and then they all leave us?' CEO: 'What happens if we don't, and they stay?'"—*Anon*.

What was the epiphany all about? Find out in the next chapter!

22.11 References

1. H.T. Laitinen, "The role of the analytical laboratory", *Anal. Chem.* **51(11)**, 1601 (1979). https://doi.org/10.1021/ac50047a600, 👆 http://bit.ly/2sgpxx6
2. K.H. Esbensen and C. Wagner, "Why we need the Theory of Sampling", *The Analytical Scientist* **21,** 30–38 (2014). https://kheconsult.com/wp-content/uploads/2017/11/WHYweneedTOS-TAS-short.pdf, 👆 bit.ly/tos1-9
3. K.H. Esbensen and C. Wagner, "Theory of Sampling (TOS) versus Measurement Uncertainty (MU)—a call for integration", *Trends Anal. Chem. (TrAC)* **57,** 93–106 (2014). https://doi.org/10.1016/j.trac.2014.02.007, 👆 bit.ly/tos16-2
4. DS 3077. *Representative Sampling—Horizontal Standard.* Danish Standards (2013). http://www.ds.dk, 👆 bit.ly/tos16-1
5. K.H. Esbensen, C. Paoletti and N. Theix (Eds), *J. AOAC Int.,* Special Guest Editor Section (SGE): Sampling for Food and Feed Materials **98(2),** 249–320 (2015). http://ingentaconnect.com/content/aoac/jaoac/2015/00000098/00000002, 👆 bit.ly/tos22-5
6. K.H. Esbensen, C. Paoletti and P. Minkkinen, "Representative sampling of large kernel lots – I. Theory of Sampling and variographic analysis", *Trends Anal Chem. (TrAC)* **32,** 154–165 (2012). https://doi.org/10.1016/j.trac.2011.09.008, 👆 bit.ly/tos8-0
7. P. Minkkinen, K.H. Esbensen and C. Paoletti, "Representative sampling of large kernel lots – II. Application to soybean sampling for GMO control", *Trends in Anal. Chem. (TrAC)* **32,** 166–178 (2012). https://doi.org/10.1016/j.trac.2011.12.001, 👆 bit.ly/tos8-2
8. K.H. Esbensen, C. Paoletti and P. Minkkinen, "Representative sampling of large kernel lots – III. General Considerations on sampling heterogeneous foods", *Trends in Anal. Chem. (TrAC)*

32, 179–184 (2012). https://doi.org/10.1016/j.trac.2011.12.002, 👆 bit.ly/tos8-3

9. K.H. Esbensen and P. Paasch-Mortensen, "Process sampling (Theory of Sampling, TOS)—the missing link in process analytical technology (PAT)", in *Process Analytical Technology, 2nd Edn*, Ed by K.A. Bakeev. Wiley, pp. 37–80 (2010). https://doi.org/10.1002/9780470689592.ch3, 👆 bit.ly/tos19-2

23 A tale of two laboratories II: resolution

This chapter completes the tale of two fictional laboratories both facing the issue: "How can the Theory of Sampling (TOS) help the commercial laboratory to improve its reputation and to increase its business"? For decades, Laboratory A has been in fierce market competition with Laboratory B, and indeed several others on the global market, which has resulted in a "healthy" business-oriented science, technology and expertise drive that has served most laboratories well (barring those who did not innovate). Both laboratories are keenly aware of the need to be in command of the TOS for all their in-house activities involving sampling, sub-sampling, mass-reduction and sample preparation. However, whereas Laboratory A has availed itself of the services of the TOS strictly within its own regimen only, as is indeed the case for most laboratories, one fine day the manager of Laboratory B had an epiphany that made her see potential advantages of applying TOS in full, involving a distinctly "beyond-the-traditional-laboratory" scope. What happened… and how did it help Laboratory B to do better in the market? Here follow three distinctly different interpretations, from which there is much to learn.

TAE: Total Analytical Error
TSE: Total Sampling Error
MU: Measurement Uncertainty

23.1 Epiphany interpretation I: knowingly closing one's eyes or not?

"A vision of a white-bearded figure carrying a tablet comes down from the mountain. The CEO can barely make out the writing, but there are the letters 'TOS' at the top … As the figure spoke of primary sampling error effects not taken proper care of, the CEO became terrified at the thought of potential implications for her laboratory ... culpability, and the ultimate terror ... litigation."

This interpretation turns decidedly serious right away... culpability, litigation... because of what? This can only relate to consequences of decisions made based on the analytical results. Which is why all commercial laboratory analytical reports carry a disclaimer, in one or many forms, the contents of which are identical, however: "The analytical results reported here, and their analytical uncertainty, pertain only to the samples delivered." For emphasis "...pertain to the *samples* delivered". This disclaimer has the clear aim to absolve the analytical laboratory of legal responsibility for any-and-all consequences of decisions made based on the assumption of representativity of the primary samples. Such decisions should be made by the client.

Most laboratories (including A and B) are undoubtedly fully aware of the risk of relatively minor effects adding to the Total Analytical Error (TAE) stemming from in-house sub-sampling, sample preparation, mass-reduction etc. in the pathway from "samples received" to analysis. All of which are very seriously addressed in any commercial laboratory enterprise whose reputation and livelihood are directly associated with the most professional command of all aspects of the science, technology and practise of *analysis*.

But the effects of the dominating primary sampling errors, if/when not dealt with still loom large in nowhere land; for which nobody is willing to take responsibility. The manager of Laboratory B realised that the consequences for believing blindly in the analytical report in this context would be borne only by the client.

23.2 Epiphany interpretation II: the economic dilemma

"The CEO of Laboratory B realised that a new business opportunity no other laboratory so far had tapped into, would be to encompass the whole process, from lot to aliquot, i.e. taking care of proper counteractions w.r.t. **both** TSE and TAE."

She felt particularly satisfied to avoid the negative statement: "Primary sampling is outside Laboratory B's responsibility", being fully aware, that by identifying this largest uncertainty component, Laboratory B would actually demonstrate its deliberate unwillingness to acknowledge the consequences hereof. Which would still have to be borne by the client alone—yet this risk, and its unquestionably dire economic consequences, is of course well known. Knowledge of these negative effects begins to look like a burden … …

"Also: Here is a possibility to increase market share! The CEO was well aware of this challenge, since no one had so far gone the whole way. And she understood the reason. Typically, clients of the laboratory only ask for the result of the aliquot analysis because they need to document the analytical results for **their** clients in turn."

A-ha, laboratories exists in a broader perspective: from-lab-to-client-to-client. As an example: analytical laboratory → consulting engineering company (e.g. responsible for environmental surveys) → regulatory authority. There are many other *similar* situations in which the entity responsible for the primary sampling is an *outsourced* entity, hired by the ultimate end-user of the analytical results. In such a case, there is typically no direct communication between the laboratory and the end-user. The market has *faith* that TAE pertains not only to the laboratory results but also the TSE part—to the degree that this "technicality" is known (which *may* well be to only a very small degree, viz. Chapter 23). The immediate client of the laboratory has no interest in correcting this, since this would only increase costs unilaterally (in order to start performing representative primary sampling). This is the traditional economic argument for continuing today's practise.

Of course, in a market-driven economy, companies (commercial laboratories are no exception), each being microeconomic ventures on their own, primarily feel responsible for their own economy. They feel that they **must** look to maximise profit before anything else. So the conventional wisdom goes in the harsh real-world of market economics.

There are two components in this aspiration: increase earnings and/or limiting costs, both defining the gap for profitability. In her dream the CEO felt very sure of being in command of this *narrow*, microeconomic competence—but, of course, just going along as usual was not really the issue any more…

"Laboratory B CEO's epiphany was a realisation that the whole package TSE + TAE was not **in demand** by the client, because the clients-of-the-client *assume* or *believe* this is included already. The CEO realised a critical need for finding tangible, compelling examples of what will happen if the TSE is ignored, specifically in terms of economic impacts for commerce, but also other, less directly tangible impacts for the public. It was felt essential to help and contribute to facilitating an efficient growing *awareness* (perhaps even public intervention) of these matters, lest 'Sampling… is gambling'!"

Laboratory B therefore also needs to address the *clients-of-the-client* in creating an **explicit demand** for a more responsible behaviour by the primary laboratory client, and thus indeed also of the laboratory itself. This will require a two-fold exercise: i) an *augmented* marketing strategy and ii) becoming involved in fostering *increased awareness* w.r.t. the TOS in general, and the dire economic effects of continuing to neglect primary sampling error effects in particular. Even in her dream of trying to break free of traditional bonds, the CEO could hear voices repeating the "board room" argument: why should Laboratory B be the one to accept larger costs for delivering the exact same type of analytical results as our competitors?

Speaking of dreams, epiphanies, nightmares—the latter often comes in the form of a *dilemma*: "I am doomed (economically) if I undertake larger costs than my competitors" and "I am doomed (morally) if I neglect the new insight that neither the client nor the client-of-the-client care one bit whether TSE is included— so long as this is **not known** by the end-user". Clearly, this is an untenable situation in any perspective. What kind of business ethics should Laboratory B adopt?

"What is common to dilemmas is **conflict**. In each case, an agent regards itself as having moral reasons to do either of two actions, but doing both actions is not possible. Ethicists have called situations like these **moral dilemmas**. The crucial features of a moral dilemma are these: the agent is required to do each of two (or more) actions; the agent can do each of the actions; but the agent cannot do both (or all) of the actions. The agent thus seems condemned to moral failure; no matter what she does, she will do something wrong (or fail to do something that she ought to do)".

23.3 Epiphany interpretation III: the moral resolution

There were some powerful insights in these two epiphanies, almost as if *written in stone*:

i) The client, and the client-of-the-client, *deserves* to know about the risk of severe economic (and other) consequences if neglecting the $TSE_{primary\ sampling}$ effects.
ii) In case this is not known to the client and/or the client-of-the-client, everybody in-the-know, Laboratory B included

of course, has a *moral obligation* to rectify this, to fill-in this factual lacuna. It cannot be right deliberately to keep one's client in the dark regarding issues that have a potential risk of severely influencing its decision making—and its bottom line adversely.

iii) WHAT will happen the day the clients of Laboratory B find out about this wilful omission?

iv) Integrity: doing what is right, regardless of whether this is known or not. Integrity is a characteristic that comes from within, based on awareness and knowledge.

The CEO realised that the integrity of Laboratory B was at stake!

The CEO realised that she would rather be CEO of a company with scientific and moral integrity, than continue to avoid a societal and moral obligation, now knowing well the adverse consequences for her company's clients and their clients in turn!

The CEO was thus now convinced that *honesty, integrity* and *transparency* must be the motto for Laboratory B's behaviour in the "analysis for sale" market. This has a necessary corollary obligation for her company. It is critically necessary to partake in a campaign for increased TOS awareness directed at everybody involved. This includes companies where sampling plays a critical role in general (**quite** a few it turned out, after just a few moments' thought) and analytical laboratories specifically (commercial as well as industrial and academic laboratories). It also includes all relevant entities in society at large, e.g. monitoring and regulatory authorities, department and governmental advisors and agencies, scientific outlets, NGOs (see Chapter 21).

As but one example of importance, EFSA (European Food Safety Agency) is charged with safeguarding the public regarding food safety and public health in all of the EU's member states. What would happen if representative sampling was **not** one of its conscious priorities? N.B. of course an entity like EFSA has a series of other major obligations and objectives, but most would suffer were not proper sampling also taken seriously. Most routine and advanced analytical characterisation of, e.g., food, feed, plants, GMOs…. are completely at the mercy of whether the relevant primary "samples" are indeed representative or not (the same argument goes for all manner of sub-sampling, sample preparation a.o. that is necessary for the complex analytical determinations needed). As the readers of this book will now know intimately, this is of imperative importance and cannot be overlooked without severe risks of adverse consequences, certainly not only of economic character, but infinitely more important, consequences for **public health** in its most broad perspective where relevant. What would happen, *hypothetically*, if the European populace one day were to find out that their public health safeguarding is not backed by absolute competence and total diligence that *also* includes sampling?

To be absolutely clear, the example of EFSA is deliberately *made up*, and *only* used here to focus the perspective, witnessed by the recently published comprehensive EFSA report specifically highlighting proper sampling.[1]

Read the EFSA report at https://doi.org/10.2903/sp.efsa.2017.EN-1226, bit.ly/tos24-1

23.4 Laboratory B's new vision and mission

The CEO laid out a new vision and mission for Laboratory B; the following obligations would henceforward be the message to its customers.

Laboratory B trusts and supports employees to take personal ownership and accountability, and learn from experiences …

Laboratory B wants to partner with its customers to enhance their productivity and performance …

Laboratory B listens to customer challenges and is actively anticipating their future requirements …

Laboratory B will do the right thing—even if it means losing a part of its earlier business ...

In the market place there would be no mercy for a company's reputation if it was revealed that the company engaged in a willing omission of disclosure and co-responsibility for the primary sampling error dominance w.r.t. the total Measurement Uncertainty ($MU_{sampling + analysis}$)—and its economic consequences. The market would not be kind in the face of a limp excuse: "But we are simply seeking to maximise our own profit—under fierce competition".

On the said "fine day" (see previous chapter), the CEO instigated a vigorous campaign for total scientific and economic responsibility and transparency. Among other initiatives she immediately made contact with appropriate TOS experts and educators in order to collaborate on this new mission: increased awareness! By doing this she was sure of minimising her own costs while maximising the benefits for clients—**and** clients-of-clients.

23.5 Can this really lead to increased commercial success?

How can one make sure that one's favourite commercial analytical laboratory, or company producing instrumental analytical equipment and "solutions", observe due diligence w.r.t. the overwhelmingly largest contributor to the **total** Measurement Uncertainty ($MU_{sampling + analysis}$)?

Easy—even a cursory visit to relevant company web sites clearly reveals whether there is the appropriate awareness, or not. The reader is encouraged to do exactly this—and observe which analytical and instrumentation company/companies instill confidence and trust in the mind of the website reader w.r.t. the so-often forgotten critical sampling issue.

The genie is out of the bottle, it is only a matter of who will be the first mover...?[7]

Will it be your laboratory?

23.6 Acknowledgements

This chapter gratefully acknowledges suggestions and input for epiphany interpretations from three colleagues. Due to the artistic licenses taken in the interest of the tale and the, perhaps, strong moral thrust developed here, they may not necessarily agree to everything laid out above and shall, for this reason, remain anonymous. But they know who they are and are thanked profusely for inspiring discussions.

23.7 References

1. K.H. Esbensen and C. Wagner, "Development and harmonisation of reliable sampling approaches for generation of data supporting GM plant risk assessment", *EFSA Supporting Publications* **14(7)**, 1226E (2017). https://doi.org/10.2903/sp.efsa.2017.EN-1226, ☞ bit.ly/tos24-1

24 Sampling commitment—and what it takes...

24.1 Historical context

The history of the World Conference of Sampling and Blending (WCSB)[1] gives a snapshot of the highly satisfactory progress in the first 10+ years since WCSB1 (2003), in which dissemination of the Theory of Sampling (TOS) has seen great strides. Reference 2 contains a plethora of earlier relevant historical references for the interested reader. WCSB1 was the inaugural world conference on sampling, and the proceedings were conceived as a comprehensive tribute to the founder of the TOS, Pierre Gy. The historical context leading up to WCSB1 can be found in Reference 3. Among Pierre Gy's last publications is a fascinating account of the history of the development of the TOS; in retrospect this turned out to be his scientific testament.[4]

WCSB: World Conference of Sampling and Blending
ISE: Incorrect Sampling Errors
CSE: Correct Sampling Errors
GP: Governing Principles
IDE: Increment Delineation Error
IEE: Increment Extraction Error
IPE: Increment Preparation Error
IWE: Increment Weighing Error
GSE: Grouping and Segregation Error
FSE: Fundamental Sampling Error

24.2 Awareness

Despite this extensive activity, there are innumerable occasions in science, technology, industry and in governing, monitoring and regulative bodies in which awareness of the need for representative sampling is still somewhat (or sometimes, very much) wanting. There are also cases on record where this knowledge is deliberately not welcomed—here we shall a.o. focus on *why* such might be the case. Introducing awareness and acknowledgement of the usefulness of applied TOS is an ongoing process that cannot be expected to be completed anytime soon (counting in decades here). There is still much work to do.

How to advance this critical awareness?

So far, where the TOS has been successfully introduced, this is mainly a result of specific drivers (dedicated individuals) who have been active within their own scientific field(s). However, now it is equally important to direct efforts to new fora in which the TOS and relevant applications have not yet been introduced. While illuminative and inspiring presentations, lectures and workshops at yearly meetings in specific science, trade and industrial fora and sectors will never fail to make a significant impact, today there is also a community which is of the persuasion that the *only thing* that counts to disseminate knowledge are **webinars**, LinkedIn entries (video snippets preferentially) and the like. History will judge which avenue fits the bill best for increasing TOS awareness. It is true, however, that systematic efforts in digital and social media are only at the very beginning. The young(er) generation(s) within the TOS community will lead the way here.

> **YouTube: a showroom of successes (a few) and failures (a lot)**
>
> One can find a plethora of more-or-less "homegrown" sampling solutions on YouTube. One is constantly surprised by the lack of proper, or any, understanding of even the most basic principles governing representative sampling on this medium, while examples of "specimenting" overflow....
>
> In this LinkedIn group there are always many "interesting" examples to peruse: https://www.linkedin.com/groups/3088030/, 👆 bit.ly/tos-24a
>
> But what is usually missing is the use of TOS principles and rules with which to assess what has been made public.
>
> Perhaps it is time that the sampling community undertook a didactic rescue mission? If this book can be an inspiration—what a joy for the author!

24.3 Minimum competence level

This chapter presents an overview of the minimum interest and comprehension necessary to acquire the scientific rationale for the TOS. This chapter also emphasises *why* the TOS is the necessary-and-sufficient framework for *any* sampling task; be this the critical primary sampling or any of the subsequent sub-sampling stages along the pathway towards a representative analytical aliquot. It is emphasised that the following applies to sampling of both stationary as well as moving lots (process sampling) of all sizes, forms and shapes.

1) **All** materials and lots in science, technology and industry are *heterogeneous*. Not knowing about heterogeneity (or not

bothering to know) is a breach of due diligence for sampling professionals, for manufacturers and for companies selling sampling equipment and solutions.

2) The primary requirement for all sampling processes, and the necessary appropriate equipment, is that of *counteracting* the effects arising from the heterogeneity met with. This is the main driving force behind all attempts to sample representatively. This demand presents a high bar for manufacturers: equipment design must be based on full TOS competence. Also, it is far from certain that a suitable sampling approach for one material will also fit the bill for another material with a significantly different heterogeneity.

3) As a minimum, it is necessary to be able to distinguish between Incorrect Sampling Errors (ISE), which lead to an inaccurate sampling process which produces the fatal *sampling bias*, and Correct Sampling Errors (CSE) which contribute to an unnecessary inflated sampling uncertainty (sampling variability). It is critically necessary to be able to distinguish between *analytical* accuracy and precision, and the *sampling* bias and precision. There is a world of difference, literally: while an analytical bias *can be* identified, quantified and thereby corrected for based on the assumption of constancy (usually a fair assumption regarding analytical methods), the sampling bias *cannot* ever be corrected because it is structurally **inconstant**.

4) The ISEs must be *eliminated* before one can get past the crippling sampling bias (or at least their effect must be substantially reduced), after which CSE must be *minimised* in order to make the sampling process sufficiently precise (reproducible). Both of these demands must be fulfilled: a representative sampling process must be both unbiased and characterised by an appropriately reduced sampling imprecision so as to become "fit-for-purpose" representative.

5) The TOS provides two facilities for estimating the effective magnitude of the uncertainty associated with any sampling process: i) the replication experiment (Chapter 9) and ii) variographic characterisation (Chapter 17). Both of these allow identification of sampling processes as fit-for-purpose representative, or which are **not** in compliance (non-representative). In the latter case, the TOS needs to be brought in competently in order to remedy the sampling stations, procedures, equipment(s) identified as inferior.

The fundamental elements of the TOS can be understood easily enough by a dedicated sampler and there is plenty of help available from today's many introductory texts. But as the reader will already know, a superior starting point is this book—and/or via dedicated workshops and courses. A first level skill can in fact be established in a remarkably short time—as short as, say, two or three days. There are no legitimate reasons to shy away from this modicum of effort in view of the goal: full comprehension of the critical understanding needed **never** to apply a particular sampling process without knowing the effective level of uncertainty that ensues. The TOS to the fore! While disregard for such a commitment would be serious enough for an individual with sampling responsibilities, picture, for example, a manufacturer selling sampling equipment or a company pitching sampling solutions *without* having demonstrated to the customer the true quality of the products and services offered. For true quality: read *proven* representativity with respect to the customer's specific materials and processes.

24.4 Vade mecum

Since 2013, there has been a *de facto* international standard, with the sole, express purpose of outlining the general principles (there are only six) and the relevant sampling unit operations (there are only four) with which to be able to address any-and-all sampling tasks—for **all** types of lots (stationary and moving lots), for **all** levels of heterogeneity (low–intermediate–high), at **all** scales and under **all** sampling conditions. This is a quite remarkable ambition level.[5–7]

Various treatises also exist dedicated to more focused sectors, e.g. the food and feed sector. "Representative sampling for food and feed materials: a critical need for food/feed safety" is a compendium, which presents the *same* universal principles and procedures for this restricted audience (Figure 24.1).[8] Be advised, however, that the treatment in here is most certainly not only valid for food and feed materials. On the contrary, this freely available publication is a general introduction of interest to all application sectors because of its universal focus.

K.H. Esbensen, C. Paoletti and N. Theix (Eds), "Special Guest Editor Section (SGE): Sampling for Food and Feed Materials", *J. AOAC Int.* **98(2)**, 249–320 (2015). 👆 http://bit.ly/tos22-5

Figure 24.1. The team behind the comprehensive introduction: "Representative sampling for food and feed materials: a critical need for food/feed safety", here caught pondering the fundamental question "how best to capture the attention of newcomers.... Left-to-right: Nancy Theix, Kim H. Esbensen, Charles Ramsey, Claudia Paoletti and Claas Wagner. Photo: Kim Esbensen

The heterogeneity of, for example, food and feed materials, of industrial commodities of all kinds and various other types of materials in technology and science in need of proper sampling is capable of being understood and counteracted by way of exactly the *same* governing principles and unit operations outlined by the TOS; this reference, therefore, forms an essential companion to the present book.

24.5 Trouble with *some* standards

It takes only a few minutes to peruse a random selection of ISO and other standards and guiding documents before one will meet a Table in which the number of increments/samples recommended is *mandated* to be proportional to the size (weight/volume) of the lot (batch, consignment) to be sampled. This is highly dubious, however, if not confined to a **very narrow**, compositionally bracketed material composition. Below follows just one simple reflection of the general lack of validity of such mandates.

Consider two lots of the same size (for the sake of argument assume large lots), but of radically different heterogeneity. One lot is of *very low* heterogeneity, in fact so low so as to come close to correspond to what in many sectors is called "uniform materials", which are defined as displaying a sampling + analysis variability for repeated sampling below 5% (in some cases below 2%). Take as an example a storage silo of refined sugar with sucrose as the analyte of interest; there is by all accounts not a formidable heterogeneity to be met here. However, were the analyte of interest a pollutant in refined sugar (or a toxic component in a similar situation for another broadly uniform lot material), i.e. a pollutant/toxic component with a minor but still adverse or dangerous concentration, clearly the distribution thereof could well constitute a quite different, more difficult to counteract heterogeneity, which will need a considerably larger number of increments. The other lot could, for example, be a run-of-the-mine broken ore (e.g. a mineralised rock with a very large difference in the proportions of valuable mineralisation, ore). Clearly it is not logical to deploy the same number of increments/samples to counteract the empirical heterogeneity met with for these two very different lot materials.

General recommendations correlated with lot size have not been considered in the light of the TOS' full understanding of the relationships between lot/material heterogeneity and the required sample mass. Small and big lots can individually be of low, intermediate or high heterogeneity; each case must be considered on its own merit. And, as always, the sampling equipment must be designed so as to correspond with TOS correctness.

For a compact introduction to these key issues, see References 5, 7 and 9.

24.6 In practice…

As an example, consider selling "sampling" equipment and/or offering sampling "solutions" **without** having completed a characterisation of the empirical lot heterogeneity met with as sampled by unit *installed* at the customer, with which to demonstrate the necessary "fit-for-purpose" representativity, based on a threshold decided upon *together* with the customer *before* installation? As major examples, the market and the literature is full of industrial hammer samplers and sampling spears (sampling thieves) that have **not** been subjected to such simple checks in practice.

Why is this so?

This issue gets all the more interesting because there are in fact a number of legitimate examples of installations of these types of sampling equipment that actually work to a sufficient

Trace concentration heterogeneity
Here are two examples from a realm in which trace concentration heterogeneity is a formidable adversary:

P. Bedard, K.H. Esbensen and S.-J. Barnes, "Empirical approach for estimating reference material heterogeneity and sample minimum test portion mass for "nuggety" precious metals (Au, Pd, Ir, Pt, Ru)", *Anal. Chem.* **88(7)**, 3504–3511 (2016). https://doi.org/10.1021/acs.analchem.5b03574, bit.ly/tos13-1

D. Desroches, L.P. Bedard, S. Lemieux and K.H. Esbensen, "Representative sampling and use of HHXRF to characterize lot and quality of quartzite at a pyrometallurgical ferrosilicon plant", *Min. Eng.* **141**, 105852 (2019). https://doi.org/10.1016/j.mineng.2019.105852, bit.ly/tos25-d

level of fit-for-purpose representativity, but this is understandably only the case for *specific lots* and materials of *demonstrated* low heterogeneity. And such a *fortuitous* case is of course **not** in any way justification for generalisation to other materials and lots!

In several practical cases this has been demonstrated beyond all doubt because the *seller* was competent and conscientious enough to be in command of the six TOS Governing Principles (GP). Even if the *buyer* should not know this, it is still the obligation of a *fair* business partner to **insist** on performing this quality check of the equipment to be sold to the customer. This is an absolutely necessary and fair business ethics demand!

The moral from the above delineates the current frontline regarding **how to** and **how much to** educate about the TOS. This concerns the central question: "Should one inform the customer in case he/she does not know enough to be able to demand proof of representativity?" Of course one should! There are also cases on record in which the customer manifestly does not want to know about this, but why and how such an attitude can develop is almost beyond comprehension, however, see section 24.7.

Most importantly: "Is **your** company, corporation, organisation, institution aware of this fundamental moral obligation?"

Is your company, corporation, organisation, institution ready to make the ultimate commitment to the TOS?

24.7 What could be argument(s) against …

What *could* be arguments against being, or becoming, TOS competent (enough) to live up to the above business ethical obligation? The author of this book is unable to conjure up **any** argument against the TOS—and never mind the likely, polemic accusation of being possessed by a gigantic *bias*! This quip aside, this author has nevertheless been exposed to a very large number of precisely such arguments during a 20-year long career within the realm of the TOS. These arguments have been presented both from academic and technological communities, but especially from many sectors from industry and commerce (see Chapter 20).

Luckily there are many counter arguments to such unwilling, ill-informed, negative attitudes towards a commitment to invoke the TOS whenever significant heterogeneity is encountered.[9,10]

By the way, how can one ascertain whether one is addressing a lot material with a significant heterogeneity, or not (hope springs eternal)? Easy—perform a replication experiment or a variographic characterisation.[10–12]

24.8   Practice, practice, practice...

In conclusion, from good (very good, excellent to brilliant) sampling to bad (ill-informed, confused, inferior, critically dangerous or fatal) "sampling" (the latter without any right to appear under such a label, the TOS has a focused term for this: worthless "specimenting"), this book ends with an example with which the reader can take a final test (without the pressure of an official exam). The test question is simple: "What's wrong with this sampler?" Even a cursory inspection will reveal several elements in blatant non-compliance with the TOS' requirements for representative sampling (Figure 24.2); for a full description see Reference 13.

Taking this type of "learning-by-lack-of-competence" to its ultimate level, please read the following abstract carefully - and please consult the corresponding Reference 14. If the reader were to "click" only one online reference in this book, it should be this overwhelmingly practical sampling showpiece by Ralph Holmes

Figure 24.2. A representative sampler? Or, a "Wheel of fortune"? The reader is encouraged to identify exactly which sampling errors are in play here from the full complement of IDE, IEE, IPE, IWE, GSE, FSE. Hint: a breakdown into primary and secondary sampling will be advantageous. The final exam will be that the reader explains, in all necessary detail, what must be done in order to rectify the disastrous procedural elements encountered here. N.B. in a way that could be explained to a complete newcomer to the realm of sampling.

in which is summed up the practical manifestations of a very large part of the teachings in this book.

> **Sample station design and operation** 👆 bit.ly/tos25-14
>
> Accurate sampling practices in the mineral industry are critical for determining the chemical, mineralogical and physical characteristics of ores and mineral products for resource evaluation and utilisation, feasibility studies, process design and optimisation, quality control, metallurgical accounting, and ultimately commercial sales. However, frequently the responsibility for sampling is entrusted to personnel who do not fully appreciate the significance and importance of collecting representative samples for analysis, and quite often everyone seems satisfied as long as some material is collected and returned to the laboratory for analysis. In the case of sample stations, cost is often the main consideration rather than sampling correctness (unbiasedness), which is unacceptable and needs to change. It is important that sampling experts are involved in the design stage at the outset to avoid structural design flaws and the subsequent need for expensive retrofits to address major and sometimes even fatal problems. Furthermore, ongoing audits of performance need to be conducted to ensure sample stations are adequately maintained and continue to conform to correct sampling principles. Provision also needs to be made for duplicate sampling to monitor the precision achieved in practice on an ongoing basis for quality assurance purposes. The examples used and commented upon here relate to one of the more difficult industry sectors with respect to correct sampling practices, material and constituent type (e.g. ores, concentrates and mineral aggregates), tonnages, process stream flow rates, and wear and tear, and as such provides the ideal showcase for the intended message which applies essentially to all technologies and industries.

24.9 The last word

This book has made its main effort to present the Theory and Practice of Sampling as a logical set of heterogeneity-related concepts, principles and practical sampling unit operations in an axiomatic manner. It is comprehensive within its restricted introductory framework, but it is, of course, far from complete w.r.t. the full foundation for which referral must be made to a series of textbooks and seminal papers, which constitute the logical next level for the interested reader. But it is the hope that the present exposé will have generated enough interest for the reader to seriously want to progress towards this next goal.

Therefore, it is appropriate to end with a select reading list of next level publications (which contain a plethora of further references). In keeping with the goal of this book, most of the references below are open access and can be easily found, while the same cannot be said for all of the selected in-depth reading literature (books a.o.), which likely have to be procured by more conventional means—but it will very much be worth the effort. Sections 24.10 and 24.11 are a portal to major elements of the more fully developed core literature. Beware of an exponentially increasing level of in-depth coverage in many of these.

Enjoy your continued TOS education!

24.10 References

1. K.H. Esbensen, *History and Achievements of the World Conference of Sampling and Blending in the Decade 2003–2013*. WCSB 6 (2013). https://intsamp.org/wp-content/uploads/2019/03/History_of_WCSB_KHE_WCSB6_proceedings.pdf, bit.ly/tos24-x
2. R.C.A. Minnitt, "The Pierre Gy Oration", *TOS Forum* **Issue 8,** 17 (2018). https://doi.org/10.1255/tosf.104, bit.ly/tos25-2

3. K.H. Esbensen, "50 years of Pierre Gy's 'Theory of Sampling'—WCSB1: a tribute", *Chemometr. Intell. Lab. Syst.* **74,** 3–6 (2004). https://doi.org/10.1016/j.chemolab.2004.06.005, bit.ly/tos25-3
4. P. Gy, "Part IV: 50 years of sampling theory—a personal history", *Chemometr. Intell. Lab. Syst.* **74,** 49–60 (2004). https://doi.org/10.1016/j.chemolab.2004.05.014, bit.ly/tos25-4
5. K.H. Esbensen and C. Wagner, "Why we need the Theory of Sampling", *Analytical Scientist* (2014). https://kheconsult.com/wp-content/uploads/2017/11/WHYweneedTOS-TAS-short.pdf, bit.ly/tos1-9
6. https://webshop.ds.dk/da-dk/standard/ds-30772013, bit.ly/tos25-6 (includes preview).
7. K.H. Esbensen and C. Wagner, "Theory of sampling (TOS) versus measurement uncertainty (MU) – A call for integration", *Trends Anal. Chem.* **57,** 93–106 (2014). https://doi.org/10.1016/j.trac.2014.02.007, bit.ly/tos16-2
8. K.H. Esbensen, C. Paoletti and N. Theix (Eds), "Special Guest Editor Section (SGE): Sampling for Food and Feed Materials", *J. AOAC Int.* **98(2),** 249–537 (2015). http://ingentaconnect.com/content/aoac/jaoac/2015/00000098/00000002, bit.ly/tos22-5
9. F.F. Pitard, *Theory of Sampling and Sampling Practice*, 3rd Edn. CRC Press, Boca Raton, Florida (2019).
10. http://kheconsult.com/a-case-for-tos/, bit.ly/tos25-10
11. K.H. Esbensen and P. Mortensen, "Process sampling (Theory of Sampling, TOS) – the missing link in Process Analytical Technology (PAT)", in *Process Analytical Technology*, 2nd Edn, Ed by K.A. Bakeev. Wiley, pp. 37–80 (2010). https://doi.org/10.1002/9780470689592.ch3, bit.ly/tos19-2
12. R.C.A. Minnitt and K.H. Esbensen, "Pierre Gy's development of the Theory of Sampling: a retrospective summary with a didactic tutorial on quantitative sampling of one-dimensional

lots", *TOS Forum* **Issue 7,** 7–19 (2017). https://doi.org/10.1255/tosf.96, 👆 bit.ly/tos16-3

13. K.H. Esbensen, "WHAT is wrong with this sampler?", *TOS Forum* **Issue 8,** 16 (2018). https://doi.org/10.1255/tosf.103, 👆 bit.ly/tos25-13

14. R.J. Holmes, "Sample station design and operation", *TOS Forum* **Issue 5,** 119 (2015). https://doi.org/10.1255/tosf.57, 👆 bit.ly/tos25-14

24.11 Further reading (a first selection)

P. Gy (1998), *Sampling for Analytical Purposes*. Wiley, Chichester (1998).

F.F. Pitard (2019), *Theory of Sampling and Sampling Practice*, 3rd Edn. CRC Press, Boca Raton, Florida (2019).

F.F. Pitard (2009), *Pierre Gy's Theory of Sampling and C.O. Ingamells' Poisson Process Approach, Pathways to Representative Sampling and Appropriate Industrial Standards*. Doctoral thesis in technologies, Aalborg University, campus Esbjerg, Niels Bohrs Vej 8, DK-67 Esbjerg, Denmark (2009). ISBN: 978-87-7606-032-9

D. François-Bongarçon and P. Gy, "The most common error in applying 'Gy's Formula' in the theory of mineral sampling and the history of the Liberation factor", in *Mineral Resource and Ore Reserve Estimation – The AusIMM Guide to Good Practice*. The Australasian Institute of Mining and Metallurgy, Melbourne, pp. 67–72 (2001).

R.J. Holmes, "Correct sampling and measurement— the foundation of accurate metallurgical accounting", *Chemometr. Intell. Lab. Sys.* **74,** 71–83 (2004). https://doi.org/10.1016/j.chemolab.2004.03.019, 👆 bit.ly/tos25-a

G. Lyman, "A brief history of sampling", *AusIMM Bulletin* 39–45 (2014).

P. Minkkinen and K.H. Esbensen, "Sampling of particulate materials with significant spatial heterogeneity - Theoretical modification of grouping and segregation factors involved with correct sampling errors: Fundamental Sampling Error and Grouping and Segregation Error", Anal. Chim. Acta **1049**, 47–64 (2019). https://doi.org/10.1016/j.aca.2018.10.056, 👆 bit.ly/tos25-b

R.C.A. Minnitt and F.F. Pitard, "Application of variography to the control of species in material process streams: an iron ore product", J. SAIMM **108(2)**, 109–122 (2008).

R.C.A. Minnitt and K.H. Esbensen, "Pierre Gy's development of the Theory of Sampling: a retrospective summary with a didactic tutorial on quantitative sampling of one-dimensional lots", TOS Forum **Issue 7**, 7–19 (2017). https://doi.org/10.1255/tosf.96, 👆 bit.ly/tos16-3

C. Ramsey, "The effect of sampling error on acceptance sampling for food safety", Proceedings of WCSB9, Beijing, May 2019. http://www.wcsb9.com/Download/, 👆 bit.ly/tos25-c

R.J. Holmes, "Best practice in sampling iron ore shipments", Proceedings WCSB9, pp. 51–62 (2019).

P.O. Minkkinen, "Cost effective estimation and monitoring of the uncertainty of chemical measurements", Proceedings WCSB9, pp. 673–685 (2019).

25 Representative sampling and society

The previous chapters have focused on the advantages the Theory of Sampling (TOS) can bring to companies, producers and manufacturers, significantly reducing costs due to inferior sampling, and maximising efficiency, logistics and profitability. Here instead, sampling is looked at from the point of view of buyers, consumers and from a broader societal perspective, exploring the economic benefits and other advantages (e.g. transparency) that can be obtained through proper sampling. This chapter explores the other side of the coin, the one linked to the ethical and moral obligations that pertain to decision-makers of responsible public and governmental bodies, which indeed should apply equally also to producers and manufacturing companies. Examples in this chapter mainly come from the food, feed, pharma, consumer products and similar sectors in society, but the potential to carry this over to the many, more traditional commodities above is obvious.

AA
GMO: Genetically Modified Organism

25.1 Sampling: from the point of view of buyers, consumers, citizens

Let us look at the role of sampling from the point of view of consumers dealing with market products which are essential in terms of both *security* and *safety*. Primary examples would be food, agricultural commodities, beverages, drugs and other medicinal products, air, soil and water quality. Here inferior sampling may not only threaten economic optimisation in the narrow production and commercial sense, but may, for example, also result in a potentially negative impact on public health. Quantitative analytical data are used daily all over the world to take important decisions which ultimately affect every single citizen; and single citizens have no other choice than to *trust* that such decisions are made on the best available basis and knowledge. The question is how, and on what basis, are decisions made regarding product and commodity safety or environmental thresholds regarding maximally allowed pollution levels? Upon reflection, there are very many such decisions that are dependent upon proper sampling… usually hidden far away in early stages of causal pathways, e.g. "from-field-to-table".

The problem is linked to the concept of "best available knowledge" for which a universal definition cannot be identified, even though it is often used to claim/guarantee quality in the interest of consumers, stakeholders and, ultimately, society at large.

However, often what is "best available"—is just not good enough.

During the last twenty years the sampling community have provided documented evidence of sampling situations where "the best available" was, and sometimes still is, insufficient. A few examples can be found in References 1–4, where the critical issue of proper sampling for GMO detection and quantification was treated in a series of papers in *Trends in Analytical Chemistry*.[2–4] In the food and feed realm, a major achievement was the 2015 special issue section of *Journal of AOAC International*: "Representative Sampling for Food and Feed Materials" presenting a compact handbook for this important societal sector,[5] complete with many consumer, user and societal viewpoints. There has also been a consistent critique of existing "sampling" sections and paragraphs in current ISO standards. Indeed, this topic remains very important: *much* of what is presented in international standards does not meet what is required to guide towards, far less guarantee, "representative sampling". The very wide spectrum of recommendations offered ranges from acceptable (not often) to "home-grown statistics" (quite often), which, although maybe correct w.r.t. the conventional statistical formulations, do not apply to the harsh reality of heterogeneity (dramatically illustrated by business case 4 in the previous chapter).[4,5]

Despite such first forays, representative sampling is not yet widely recognised as one of the key tools needed to ensure that the quantitative analytical data used to take subject-matter decisions truly are the "best available". As responsible sampling scientists and practitioners, we must be realistic and accept that, on the present basis, claims identifying the TOS as the *only* framework for correct sampling may not always be understood; there is much more work to be done before significant impacts on the general population can be achieved. A newly released report dealing with a topic that runs parallel to the present (indeed it overlaps significantly: proper sampling is a critical prerequisite to circular economy)[6] shows this compellingly. Another, "Barriers to the circular economy: evidence from the European Union" by Kircherr *et al.*[7] reveals the complexity and immensity of this kind of awareness and educational endeavours.

K.H. Esbensen and C. Velis, "Editorial: Transition to circular economy requires reliable statistical quantification and control of uncertainty and variability in waste", *Waste Manage. Res.* (Dec. 2016). https://kheconsult.com/wp-content/uploads/2017/11/WMR680911_CIRCULAR.pdf, http://bit.ly/tos22-6

Sampling champions feel a moral obligation to find innovative ways and means to incorporate representative sampling as a key criterion for any quality statement, ensuring a step-forward in the correct application of scientific knowledge to practice.

So, two basic questions arise:
1) **Who** should decide when "the best available" is indeed the best?
2) **How** can we convince stakeholders and citizens that correct sampling is a necessary pre-requisite, among others, to ensure security and safety of the relevant products and services essential for society?

25.1.1 Addressing both questions

1) Normally consumers decide quality, but this rule is difficult to apply when the quality under discussion is not the one of the final product(s), but rather of the *process* used to manufacture, process or deliver products (or services). Suddenly, quality becomes *invisible* for the consumer. This is why *individual citizens* can only trust that market decisions taken for essential products are made on the best available knowledge and must be happy (if not happy... at least willing) to pay taxes so that public control systems have sufficient resources to protect them! This admittedly oversimplified scenario is meant to illustrate the ethical responsibility that regulatory science bears towards society, a complex responsibility. But when one accepts this logic, we can easily answer the first question, **who** should decide when "the best available" is indeed the best? Only those having sufficient experience and knowledge of the process leading to a product can decide if "the best available" information is sufficient.

If we project these considerations to sampling, it becomes clear that the quality of the sampling used in the decision-making regarding products essential for society cannot be assessed by the individual final consumer. Assessment of sampling quality relies on the professional integrity, expertise and objectivity of those controlling the production process. This completely changes the frame within which sampling problems are addressed and resolved, making it incomparable to that faced by the TOS consultants working in the commercial realms, where the quality of their work is assessed directly by their clients. The part of the TOS community interested in engaging in sampling of societally essential products must be fully aware of these additional difficulties and responsibilities, which can frustrate (hopefully only temporarily) even the most motivated and determined sampling expert.

2) However, even if society in its role as final consumer of essential goods cannot monitor the quality of processes, it should be educated and aware (enough) to fully appreciate the practical relevance and implications that representative sampling has, even if invisible to its final consumers. Here is a fact simple enough to be intuitively understandable **by all**: if sampling is not representative it is futile, indeed useless, to analyse the ensuing "samples", because it has no meaning to produce such analytical results without a clear provenance; the sampling + analytical uncertainty becomes totally unknown. This issue has been well illustrated in previous chapters and has been explained many times in various public fora. So much so that, gradually, various international normative documents now do mention that "good sampling *should be* representative". True, this is a much milder and timid version of "non-representative sampling is useless", but whether we like it or not, this is currently the only reward the sampling community has received for some 15–20 years of hard work. Now is the time to explore new strategies to speed up progress and ensure that representative sampling becomes a central element in the list of the essential quality criteria. However, exactly **how to do so** requires careful thinking, because it will unavoidably entail identification and correction of deficiencies in current practices, which of course is never popular. Examples of the first steps in this direction could be References 8 and 9, against which there is non-trivial resistance. These issues were plentifully illustrated in Chapter 20: "Sampling—*Pro et Contra*".

25.2 The way forward: some proposals

First, we should better *substantiate* the claim that the TOS is the **only** sampling framework universally applicable to any type of material and heterogeneity. We should demonstrate, with empirical evidence, that this is in fact the case.

Encore
The reader is referred to the margin box on page 19 dealing with "only one" Theory of Sampling (TOS)

Figure 25.1. "Where it all begins." The dominating errors behind the final analytical uncertainty are always largest at the primary sampling step. Here soy beans are off-loaded from a cargo ship's holds. It is decidedly not a trivial issue how to sample this type of lot in a documented representative fashion—professional TOS expertise is needed. By now the reader should be close, or maybe already able, to form a professional opinion: intersect the 1-D lot stream that commences with the grain elevator shown followed by a subsequent conveyor belt.

The KeLDA project[2-4] did so ten years ago (Figure 25.1), but no other examples of similar dimension (outside the mining and minerals processing sectors) have been produced since. No misunderstanding regarding the mining sector: from here comes the evergreen "How much the TOS saves you in monetary terms" publication paper *par excellence* by Carrasco, Carrasco and Jada: "The economic impact of correct sampling and analysis practices in the copper mining industry",[10] which was also summarised in Chapter 21.

The sampling community has provided seminal textbooks and many excellent scientific papers explaining with various degrees of complexity and comprehensibility the mathematics upon which the TOS is rooted, and where the TOS is currently progressing technically. The series of Proceedings, from the bi-annual World Conferences on Sampling and Blending, in which applications to a steadily broadening societal field are presented, constitutes a further, highly significant progress in this context.

IPGSA homepages\WCSB Proceedings:
https://intsamp.org/proceedings/, 👆 http://bit.ly/tos20-3

Still the sampling community must continue to make extra efforts to put itself on the side of society, where intimidating mathematical formulae are respected, but only occasionally understood (and only very rarely actually read), but where practical/direct examples are seen as the primary evidence that allows the light to be seen in what is perceived as an intricate forest of technical and scientific complexity.

History teaches that significant, mass-scale changes in attitude towards scientific innovation have only taken place when the triggering explanations were simple, clear and self-evident.

The TOS community has not yet found a fully comprehensive, winning way to achieve this, but the traditional emphasis on beneficial as well as adverse economic consequences resulting from involving proper TOS, or not, is covered extensively in Chapter 21. Below follows a few views on what *can* perhaps also be done to trigger increased societal attention to sampling.

25.3   Beyond traditional application fields

The use of fortified foods, food supplements and "functional foods" is on the rise. This may result in a higher intake of nutrient substances, which *could* turn into a concern if intake levels become sufficiently high to induce adverse effects. Nutrients, in contrast to contaminants, are essential for human/animal health and have their positive nutritional effects within specific concentration ranges, governed by homeostatic mechanisms in the human/animal body. Adverse health effects may occur due to over-consumption or may lead to deficiency symptoms in case of under-consumption. Therefore, upper intake levels of nutrients from food sources by humans/animals that do not induce adverse health effects and minimal required intake levels should be identified in order to avoid such effects. Obviously, proper sampling methods applied at various stages of production and processing of these foods are needed to be able to correctly determine actual intake levels of nutrients by humans and compare these with the established upper safety limits and minimal required intake levels.

Another well-known fact is the increased spread of *pathogens* in the food production chain, presumably due to globalisation of trade and to the migration of people.[11] New pathogenic micro-organisms have been detected and characterised, as well as an increase in antibiotic-resistant bacteria, presumably due to massive (over-)use of antibiotics for human therapy. Ingestion of pathogens or their toxins may induce a variety of diseases in humans/animals, ranging from acute illness like diarrhoea to many chronic diseases and even death. Specific guidance for risk assessment of microbial food and feed contamination has been developed[12] and the importance of the dynamics of microbiological growth, survival and the (rapid) transfer of micro-organisms throughout the food production chain in many types of foods, raw or processed, and further spread in the environment has been underlined. Exposure assessment is of critical importance for risk assessment and consequently also for definition of suitable sampling plans, that take into account the specific distributional characteristics of microbial populations and of their spreading dynamics. These issues are of the utmost importance to allow an effective safety evaluation of food and feed commodities.

Consumption and demand for *niche and brand* products, e.g. mono-cultural products, extra-virgin olive oil, mozzarella cheese, designed to capture the interest of an elite portion of consumers, is also increasing, at least in more developed and wealthy countries. In such cases, proper sampling may raise interest in **both** producers and society. For producers, correct sampling may facilitate the conquest of a portion of the market at the global level, ensuring/proving specific quality standards of unique products. Indeed, some producers are aggregating into consortia with the objective of facilitating their business. For society, the same holds true: representative sampling becomes a tool to ensure that the final niche-product on the market, possibly at a higher price to cover the specific production costs, indeed has the compositional, organoleptic and nutritional properties it is claimed to have. Here proper sampling benefits both sides of the traditional producer–consumer issue equally.

The TOS community needs to accept that sampling is up against a series of inherent difficulties linked to the nature of products in a wide societal sense, in particular beyond the TOS' traditional target fields of mining, minerals processing and cement. The great diversity in food and feed sources, commodities and the different kinds and degrees of food/water contaminations are just a few examples, focussed on the difficulty for society to directly verify the quality of the production processes involved. There are assuredly many other societal sectors with similar issues.

Finally, here is a problem that only a few want to entertain today: the sampling frameworks currently used for quality assessments too often rely on *specific* statistical distributional assumptions (i.e. "homogeneous distributions" of compounds/test materials, "assumed" normal distributions), but which are very nearly *never* **verified** in practice, as current protocols do not even stipulate characterisation of inherent heterogeneity patterns stemming from the specific properties of the targeted materials[9]. Moreover, current quality assessment protocols do not provide estimates of the risk associated with the sampling surveys *themselves*, nor do they address the uncertainties associated with spatially irregular distributions (material distributional heterogeneity).[5,8]

25.4   Conclusions

Above, it has been underlined that representative sampling is **key** in order to reduce the possibilities of either mis-estimation of actual exposure levels for humans and animals or, worse, underestimating the risks for consumers to exceed tolerable intake levels. We need to prove this correct framework understanding with real data and demonstrations. Returning to the introduction of this chapter, this demand regarding public health risk assessments has a significant potential to carry over to the many other materials and commodity types addressed in this book.

When one goes to a wine cellar or a wine store, one tastes (*samples*) the wine before buying it: you do not ask the seller **if** the wine is good. The TOS community cannot expect to be listened to by non-experts and the general populace if it cannot also document its claims with **facts**, compelling visual, graphic and quantitative facts, including the economic be-all/end-all: costs or gains? This calls for superior examples and resources to support well-thought out research and demonstration projects with the same purposes. One is reminded of the data insight brilliance by Professor Hans Rosling (1924–2017).[13]

The present book is a deliberate attempt to offer a conceptual "TOS factfulness" bearing in mind all the possible error effects remaining in "data" if not properly dealt with.

Hans Rosling 1948–2017
Author of the inspiring book *Factfulness*.

https://en.wikipedia.org/wiki/Hans_Rosling

25.5 References

1. K.H. Esbensen, F. Pitard and C. Paoletti, "Sampling errors undermine valid genetically modified organism (GMO) analysis", *TOS forum* **Issue 1,** 25–26 (2013). https://doi.org/10.1255/tosf.9, bit.ly/tos22-1
2. K.H. Esbensen, C. Paoletti and P. Minkkinen, "Representative sampling of large kernel lots – I. Theory of Sampling and variographic analysis", *Trends Anal. Chem.* **32,** 154–165 (2012). https://doi.org/10.1016/j.trac.2011.09.008, bit.ly/tos8-0
3. P. Minkkinen, K.H. Esbensen and C. Paoletti, "Representative sampling of large kernel lots – II. Application to soybean sampling for GMO control", *Trends Anal. Chem.* **32,** 166–178 (2012). https://doi.org/10.1016/j.trac.2011.12.001, bit.ly/tos8-2
4. K.H. Esbensen, C. Paoletti and P. Minkkinen, "Representative sampling of large kernel lots – III. General Considerations on sampling heterogeneous foods", *Trends Anal. Chem.* **32,** 179–184 (2012). https://doi.org/10.1016/j.trac.2011.12.002, bit.ly/tos8-3
5. K.H. Esbensen, C. Paoletti and N. Theix (Eds), "Special Guest Editor Section (SGE): Sampling for Food and Feed Materials", *J. AOAC Int.* **98(2),** 249–537 (2015). http://ingentaconnect.com/content/aoac/jaoac/2015/00000098/00000002, bit.ly/tos22-5
6. K.H. Esbensen and C. Velis, "Editorial: Transition to circular economy requires reliable statistical quantification and control of uncertainty and variability in waste", *Waste Manage. Res.* (Dec. 2016). https://kheconsult.com/wp-content/uploads/2017/11/WMR680911_CIRCULAR.pdf, bit.ly/tos22-6
7. J. Kirchherr, L. Piscicelli, R. Bour, E. Kostense-Smit, J. Muller, A. Huibrechtse-Truijens and M. Hekkert, "Barriers to the circular economy: evidence from the European Union (EU)",

Ecol. Econ. **150,** 264–272 (2018). https://doi.org/10.1016/j.ecolecon.2018.04.028, bit.ly/tos22-7
8. H.A. Kuiper and C. Paoletti, "Food and feed safety assessment: the importance of proper sampling", *J. AOAC Int.* **98,** 252–258 (2015). https://doi.org/10.5740/jaoacint.15-007, bit.ly/tos22-8
9. C. Paoletti and K.H. Esbensen, "Distributional assumptions in food and feed commodities – development of fit-for-purpose sampling protocols", *J. AOAC. Int.* **98,** 295–300 (2015). https://doi.org/10.5740/jaoacint.14-250, bit.ly/tos22-9
10. P. Carasco, P. Carasco and E. Jara, "The economic impact of correct sampling and analysis practices in the copper mining industry", in Special Issue: 50 years of Pierre Gy's Theory of Sampling. Proceedings: First World Conference on Sampling and Blending (WCSB1), Ed by K.H. Esbensen and P. Minkkinen, *Chemometr. Intel. Lab. Syst.* **74(1),** 209–213 (2004). https://doi.org/10.1016/j.chemolab.2004.04.013, bit.ly/tos21-1
11. L. Saker, L. Kelly, B., Cannito, A., Gilmore and D.H. Campbell-Lendrum, *Globalization and Infectious Diseases: A Review of the Linkages.* World Health Organization, Geneva (2000).
12. Report of a Joint FAO/WHO Consultation entitled *Principles and Guidelines for Incorporating Microbiological Risk Assessment in the Development of Food Safety Standards, Guidelines and Related Texts.* FAO and WHO (2002).
13. H. Rosling, *Factfulness.* Sceptre Publishing (2018). ISBN 978-1-473-63746-7

26 Epilogue: what's next?

Congratulations: you have just passed the biggest hurdle, by far, in the realm of the *Theory and Practice of Representative Sampling*, and by reading this book you have passed it with flying colours! You have now *mastered* the core elements of the TOS: six Governing Principles (GP), four Sampling Unit Operations (SUO) and you have formulated your own mind map of the crucial sampling error management rules: how to understand, identify and, most importantly, how to *deal with* no less than eight CSE and ISE. The singular lesson from the preceding 25 chapters was that the first item on any sampling agenda is always to eliminate (or reduce as much as possible w.r.t. a fit-for-purpose representativity criterion) the Incorrect Sampling Errors, as these otherwise cause a fatal sampling bias invading your sampling protocol! And you now know that you can do absolutely nothing about that after the fact—**only** preventive TOS knowledge can help you. Which is why the role of designing sampling procedures correctly from the start becomes imperative.

There you have it—the essence of this book—although we had to use 323 pages to make sure you feel confident as a competent sampler, who is now able to see behind quite a number of apparently bewildering manifestations of potential effects (CSE, ISE). The author of this book feels confident enough to let you start on your own sampling journey! But it is with sampling, exactly like it is with driving a car. After you are issued with your first driver's licence, nobody expects you (least of all yourself) to be an expert driver right off—you now need a great deal of practical experience. So, as soon as you have passed one goal, you discover that the race is not over, but continues… albeit on a higher level.

GP: Governing Principle
SUO: Sampling Unit Operation
CSE: Correct Sampling Errors
ISE: Incorrect Sampling Errors
PTE: Process Trend Error
PPE: Process Periodicity Error
TSE: Total Sampling Error

But wait a minute: eight sampling errors? You are now very familiar with the six basic sampling errors, but not eight. You were hardly offered more than an initial *naming* of the two process sampling errors, the Process Trend Error (PTE) and the Process Periodicity Error (PPE), but you were *promised* that this would not hamper you in getting that crucial next-level experience. This is true—but this small (PTE, PPE) issue does point to perhaps the most important area of applied TOS, to the arena of *process sampling*. Which, although it was properly *introduced* to the reader, was certainly not treated in full. This is where the forefront of sampling has been in all of the TOS's 60 years of existence. So why was there not more of this subject in the book's 26 chapters?

Because it is imperative to master all the concepts, principles, rules and to have built up a wealth of experience regarding sampling of *stationary lots* **before** one can delve more fully into process sampling in any meaningful fashion. So, there is still a lot to learn, but it is all at the next level from where you stand today. You may be relieved to hear that in mastering all there is to know about the sampling of stationary lots, you have already acquired at least 75 % of the skills needed to extend your mastery to also include sampling of moving, dynamic lots.

So, this is "What's next?" for you. This is why you will hopefully be pleased to know that a Book II is in the offing, conceived exactly like the one you have just completed: *Introduction to the Theory and Practice of Representative Process Sampling* is slated for publication in 2021 (hopefully ready for launching at the 10th anniversary World Conference on Sampling and Blending, WCSB10, in Kristiansand, Norway, June 2021.

Here is a sneak-preview of the contents of Book II.

1) Applications, applications, applications… (a deliberate echo of the "Location, location, location" dictum in real estate marketing). There is nothing more important for cementing your newly acquired TOS skills than further exposure to the

Epilogue: what's next?

sampling demands from the real world and its plethora of significantly heterogeneous materials! Applications will cover both more stationary lot sampling cases, but will, of course, feature process sampling cases prominently.

2) *Some* theory will also now be appropriate, guiding the reader towards a deeper understanding of the TOS, to be presented where relevant. N.B. we will even venture to add a certain measure of the TOS' original mathematical language, but, as always, the reader decides where to invest her/his efforts. However, we will spare no efforts ourselves to make these presentations so interesting and compelling that the reader will have no real choice!

3) It will be a pleasure to prepare an assorted compilation of compelling didactic examples with which to fortify the teachings of both Books I and II—see also item 6 below.

4) Book II will also present a bonus to the reader, by now hopefully an ardent sampler for whom the legacy of untreated ISE is only to be sneered at: the VARIO freeware. In addition to allowing the reader to make the basic, yet always powerful variograms already implemented as free software in the DS 3077 standard (but here limited to 100 samples), VARIO will now allow the TOS-competent sampler to analyse an *unlimited* number of samples, and will expect the sampler actively to direct how to estimate the all-important nugget effect, $V(0)$, which will result in VARIO's main feature, comprehensive estimation of the TSE associated with the user's initial choice of *sampling mode* (regular, random, stratified random). It goes almost without saying that such a powerful weapon demands a fairly high skill level from the user—hence no VARIO software before Book II.

5) Consequently, Book II will include many process sampling examples and case histories in which proper, high-level use of variographic analysis is central.

> For the impatient reader, *parts* of Book II are already available. A major part of Book II will be edited, and augmented, versions of what is continually being presented in the Sampling Columns in *Spectroscopy Europe/Asia* (https://www.spectroscopyeurope.com/sampling, bit.ly/tos20-2). The reader is encouraged to consult this source; the first installments (future chapters in Book II) are already on-line. Enjoy!

Whether you have worked your way through this book as part of a TOS course, or via solitary reading, be assured you are not alone on your continued journey. You are always welcome to contact the author. No sampling issue is too small, or too large—it is always of interest: www.kheconsult.com

6) Hopefully interesting to many readers, Book II also plans to include examples, case histories and didactic demonstrations from the *readers themselves*. This could be **YOU**! I have found it a pleasant experience from conducting many TOS courses and workshops, that often the most interesting examples come *from the audience*... upon reflection for the good reason that this is the depository of as yet undiscovered problems and contexts in which sampling plays an integral role. The reader is encouraged to consider *participating* in this joint venture. Both the Sampling Columns in *Spectroscopy Europa/Asia*, as well as *TOS Forum* (https://www.impopen.com/tos-forum, 👆 bit.ly/tos21-a) would welcome your contributions!

> **Judgement time**
>
> At last, now the reader should be able to pass the final judgement for her/himself: "What works best to capture the attention of anyone who has not been introduced to the issue of "WHY sampling?" before:
>
> i) The economic argument(s)? "You'll lose a lot of money, if you don't know about the TOS!"; or the technical argument:
>
> ii) "Your decisions are only as good as your samples—are they representative, and what does that mean?"
>
> This book has ventured to accomplish what surely looked as Mission Impossible at the start: HOW can something as "complex" as the TOS be presented so as to be fully understood, and equally well, by all stakeholders: process technicians, process engineers, scientists, laboratory staff, investors, management (CEO, CTO, CFO)... to name but a few?
>
> The reader is now fully equipped to make up her/his own mind—which is one of the most satisfactory aspects of having written this book!

> **Testimony**
>
> Understanding what sampling variation is, and how it is estimated, has been a "light-bulb" moment for our analysts after having been introduced to the Theory of Sampling (TOS) principles. So often we have had a situation where analytical work and results can be verified, but our customer still insists it doesn't meet expectations. Short of driving the poor analyst crazy with re-work tasks, which usually only produces the same "incorrect result", I now have an avenue of action that allows us to guide the customer and analysts to the path on how to focus on only taking representative samples. This is decidedly more welcome than always having to hear: "Take the sample back to the lab—repeat the analysis".
>
> Much time is spent determining the combined total uncertainty for specific analytical methods under validation, however, very little attention is given to the preceding sampling errors and the challenges heterogeneity poses to this issue. I now know that sampling errors dominate over their analytical cousins. Also, using variographic characterisation as a quality control tool for process and measurement system monitoring is a very powerful technique that could help process controllers explain the sources of real process variations that occur on their product lines instead of simply following through by blaming the analytical lab. I found that the new international standard DS 3077 (2013) and in particular its use of illustrations and industrial examples captured the true complexity of the principal types of Sampling Errors and helped to conceptualise the TOS principles in a strikingly visual way, making it easier for a typical chemical analyst to relate to the scenarios involved before analysis. After all, we have to isolate the absolutely smallest aliquot for analysis—as demanded by highly sophisticated analytical instrumentation. It is, therefore, highly surprising that the one area of greatest error affecting analysts' results is the same topic largely ignored in Analytical Chemistry/Science Training programmes, again the sampling errors. This gives rise to "brilliant" analytical results, i.e. extremely precise results, but for non-representative samples for which accuracy with respect to the lot is not accounted for. In fact the accuracy of the analytical results with reference to the original lot is completely without control—and one cannot even estimate the magnitude of the sampling bias incurred (because it is inconstant, as is another insight provided by TOS). This makes for a very unsure analytical laboratory. After this course I wonder how many questionable results have been released by laboratories all over the world over many, many decades—and the revelations brought about by TOS are still not known!
>
> *Dr Melissa C. Gouws, InnoVenton Analytical, Port Elizabeth, South Africa*

"All sampling procedures invoked to secure primary samples (as well as all sub-sampling operations needed to produce the analytical aliquot), whether by buyer, seller or an arbitration agency, shall be compliant with the principles of representative sampling as laid out by the Theory of Sampling (TOS), as codified in the standard DS 3077 (2013). All sampling procedures involved must be adequately and fully documented."

KHEConsulting

Comprehensive TOS course
Theory and Pratice of Sampling of Heterogeneous Materials and Processes

This course presents The Theory of Sampling (TOS) in a novel systematic way presenting six Governing Principles (GP) for guiding optimal application of four Sampling Unit Operations (SUO). The course hallmark is its practical approach with abundant examples and facilitating case histories, focusing more on an overview framework understanding and less on the abundant mathematical background details. Starting out by covering sampling from stationary lots as a means of 'learning the ropes', via the critical issue of proper sampling in the laboratory, the course concludes with a comprehensive focus on industrial and technological process sampling. This course facilitates effective competence and know-how transfer to scientists, technicians, engineers, process operators, laboratory personal and managers.

Course brief

A set of six Governing Principles (GP) and four Sampling Unit Operations (SUO) cover all practical aspects of sampling and provides a unifying framework for all stakeholders, scientists (academic staff, Ph.D. students), technologists, process engineers, process operators, laboratory and industry personnel, presenting the necessary competence needed to guarantee that all primary sampling, splitting, sub-sampling and sample preparation stages are fully representative (procedures, equipment, maintenance). There is a special part on "proper sampling in the analytical laboratory".

KHEC is a world- leading educational and competence building international consultancy for representative sampling in all technological and industrial sectors. At all stages, from primary sampling, sub-sampling and sample handling in the lot-to- aliquot pathway, KHEC has a professional obligation to act on behalf of the client's interest regarding all sampling matters[1].

In close cooperation with customers, KHEC also develops sampling solutions for non-standard cases and applications. KHEC is a responsible partner from procedures to verification of existing and new installations. All design and procedure suggestions follow the guidelines given by TOS offering the potential for realising optimal fit-for-purpose solutions for each customer's specific needs. KHEC also offers audits of complete sampling systems.

1. "A Tale of two Laboratories I & II" https://www.spectroscopyeurope.com/sampling

KHEConsulting

This 3-day course provides a complete introduction to the TOS for stationary lots and for dynamic lots. The course also makes the critical connection to process engineering, Process Analytical Technologies (PAT) and multivariate data analysis (chemometrics).

Course goal

The sampling bias has a fundamentally different nature than the analytical bias, sadly negating all attempts of 'sampling bias-correction'. This is the greatest surprise provided by this course. The test of all sampling systems is whether the sampling bias has been eliminated or not TOS provides a set of practical ways to achieve "sampling correctness" (unbiasedness) by informed understanding, design and application of the relevant principles in the form of an unbiased sampling process and the relevant equipment.

The course gives full insight into how to guarantee that all primary sampling, and subsequent sub-sampling (splitting) and sample preparation before analysis is documentable as representative (procedures, equipment, maintenance). After the critical primary sampling step, correct (unbiased), mass reduction (splitting) in the subsequent sub-sampling in the laboratory also needs to be 100% compliant with TOS in order to ensure valid analytical results. It is often unknown, or is willfully neglected, that the Total Sampling Error (TSE) is by far the dominating contribution to the total Measurement Uncertainty (MU), typically 10-25 X larger than the Total Analytical Error (TAE). This is a fact neglected by the discipline of Measurement Uncertainty (MU); TOS provides a seamless integrated solution.

Representative sampling is the critical success factor for achieving optimal analytical accuracy and precision - as needed for reliable decision making in science, technology and industry. All steps in the lot-to-aliquot pathway shall be compliant with DS3077 (2013)[1], today's de facto international standard for representative sampling.

There is significant added value in a common TOS competence in order to maximise clients' and customers' potential only to use representative solutions and equipment. Collaboration is furthered through general and dedicated in-house seminars and courses on "Theory and Practice of Representative Sampling, TOS".

1. DS3077 (2013) "Representative Sampling – Horizontal Standard"   www.ds.dk

KHEConsulting

This course provides attendees a comprehensive overview of the Theory of Sampling (TOS) for stationary lots as well as process lots and in the laboratory, including powerful facilities with which to characterise lot heterogeneity, called Replication Experiments (RE) and variographic characterization which allows for improved process understanding and total process system measurement system

Sampling bias – behind the myth

The sampling bias has a fundamentally different nature than the analytical bias, negating all attempts of 'sampling bias-correction'. This is perhaps the greatest surprise provided by this course. Instead TOS provides a set of practical ways to achieve "sampling correctness" (unbiasedness) by informed understanding, design and application of the generic sampling process and the relevant equipment. The course overview gives full insight into how to guarantee that all primary sampling, and subsequent sub-sampling (splitting) and sample preparation before analysis is documentable as representative (procedures, equipment, maintenance).

After the critical primary sampling step, correct (unbiased) mass reduction (splitting) in the subsequent sub-sampling in the laboratory also needs to be 100% compliant with TOS in order to ensure valid analysis analytical results. It is often unknown, or is willfully neglected, that the Total Sampling Error (TSE) is by far the dominating contribution to the total Measurement Uncertainty (MU), typically 10-25 X larger than the Total Analytical Error (TAE). This is a fact neglected by the discipline of Measurement Uncertainty (MU); TOS provides a seamless integrated solution.

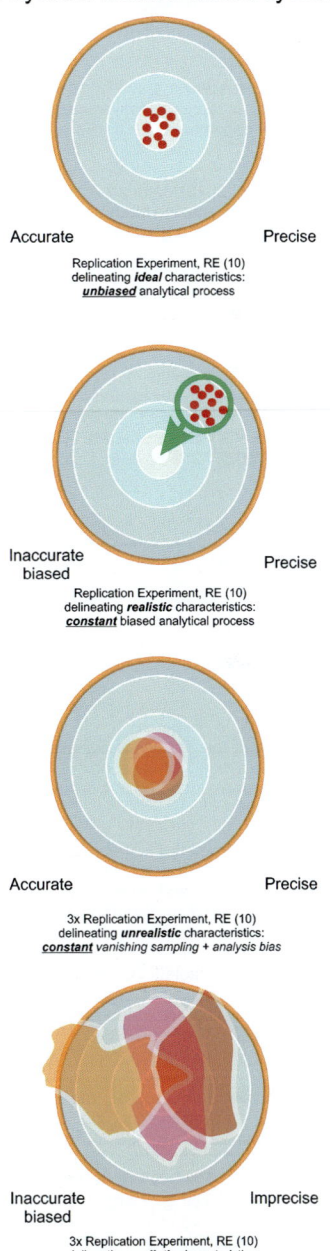

Accurate — Precise
Replication Experiment, RE (10) delineating *ideal* characteristics:
unbiased analytical process

Inaccurate biased — Precise
Replication Experiment, RE (10) delineating *realistic* characteristics:
constant biased analytical process

Accurate — Precise
3x Replication Experiment, RE (10) delineating *unrealistic* characteristics:
constant vanishing sampling + analysis bias

Inaccurate biased — Imprecise
3x Replication Experiment, RE (10) delineating *realistic* characteristics:
inconstant non-vanishing sampling + analysis bias

KHEConsulting

Course literature

The course includes a comprehensive literature documentation, including the world's first standard dedicated exclusively to representative sampling, DS3077 (2013).

DS 3077 (2013) "Representative Sampling – Horizontal Standard" https://webshop.ds.dk/da-dk/standard/ds-30772013

Esbensen, K.H. & Julius, L. (2013) "DS 3077 Horizontal—a new standard for represent tative sampling. Design, history and acknowledgements", TOS Forum 1, p. 19-22 doi: 10.1255/tosf.7

Esbensen, K.H. (2015) Materials Properties: Heterogeneity and Appropriate Sampling Modes. J. AOAC Int. vol. 98, pp. 269-274. http://dx.doi.org/10.5740/jaoacint.14-234

Esbensen, K.H., Wagner, C. (2014). Theory of Sampling (TOS) vs. Measurement Uncertainty (MU) – a call for integration. Trends in Analytical Chemistry (TrAC) 57, 93-106

Esbensen, K.H. & Julius, L.P. (2009). Representative sampling, data quality, validation – a necessary trinity in chemometrics. in Brown, S, Tauler, R, Walczak,R (Eds.) COMPREHENSIVE CHEMOMETRICS, Wiley Major Reference Works, vol. 4, pp.1-20. Oxford: Elsevier

Petersen, L, C. Dahl, K.H. Esbensen (2004). Representative mass reduction in sampling – a critical survey of techniques and hardware. Chemometrics and Intelligent Laboratory Systems, vol. 74, Issue 1, p. 95-114

Esbensen, K.H. & Mortensen, P. (2010). Process Sampling (Theory of Sampling, TOS) – the Missing Link in Process Analytical Technology (PAT). in Bakeev, K. A. (Ed.) Process Analytical Technology. 2.nd Edition. pp. 37-80. Wiley. ISBN 978-0-470-72207-7

Minnitt, R.C.A. & Esbensen, K.H. (2017) Pierre Gy's development of the Theory of Sampling: a retrospective summary with a didactic tutorial on quantitative sampling of one-dimensional lots. TOS Forum 7, p. 7-19. doi: 10.1255/tosf.96

Esbensen, K.H, Paoletti, C, Theix, N. (2015) (Eds) Journal AOAC International, Special Guest Editor Section (SGE): Sampling for Food and Feed Materials. pp. 249-320 http://ingentaconnect.com/content/aoac/jaoac/2015/0000 0098/00000002

Esbensen, K.H., Paoletti, C. & Minkkinen, P. (2012). Representative sampling of large kernel lots – I. Theory of Sampling and variographic analysis. Trends in Analytical Chemistry (TrAC), 32 pp.154-165

Minkkinen, P., Esbensen, K.H. & Paoletti, C. (2012). Representative sampling of large kernel lots – II. Application to soybean sampling for GMO control. Trends in Analytical Chemistry (TrAC), 32 pp. 166-178

Esbensen, K.H., Paoletti, C. & Minkkinen, P. (2012). Representative sampling of large kernel lots – III. General considerations on sampling heterogeneous materials. Trends in Analytical Chemistry (TrAC), 32 pp. 179-184

https://www.impopen.com/sampling

Course presenter

Professor Kim H. Esbensen,
owner
KHE Consulting (KHEC),
Copenhagen

This course is offered as a 2 + 1 day comprehensive course; the venue is either at KHEC's domicile, Copenhagen or at client's location (in-house course).

The course is also offered in a compact 1-day overview format.

Contact

☎ **Call me**
  +45 20 21 45 25
✉ **Write me**
  khe.consult@gmail.com
🏠 **Webadress**
  www.kheconsult.com